ELECTRICAL ENGINEERING DEVELOPMENTS

APPLICATIONS OF VARIOUS FUZZY SLIDING MODE CONTROLLERS IN INDUCTION MOTOR DRIVES

ELECTRICAL ENGINEERING DEVELOPMENTS

Additional books in this series can be found on Nova's website under the Series tab.

Additional e-books in this series can be found on Nova's website under the eBooks tab.

ELECTRICAL ENGINEERING DEVELOPMENTS

APPLICATIONS OF VARIOUS FUZZY SLIDING MODE CONTROLLERS IN INDUCTION MOTOR DRIVES

ALI SAGHAFINIA, PHD

nova publishers
New York

Library of Congress Cataloging-in-Publication Data

Names: Saghafinia, Ali, editor.
Title: Applications of various fuzzy sliding mode controllers in induction motor drives / [edited by] Ali Saghafinia (Electrical Engineering Department, Majlesi Branch, Islamic Azad University, Esfahan, Iran).
Description: Hauppauge, New York : Nova Science Publishers, Inc., [2016] | Series: Electrical engineering developments | Includes bibliographical references and index.
Identifiers: LCCN 2016017082 (print) | LCCN 2016022579 (ebook) | ISBN 9781634851794 (hardcover) | ISBN 9781634852340 (ebook) | ISBN 9781634852340 ()
Subjects: LCSH: Electric driving--Automatic control. | Electric motors, Induction. | Electric controllers. | Fuzzy automata.
Classification: LCC TK2785 .A75 2016 (print) | LCC TK2785 (ebook) | DDC 621.46--dc23
LC record available at https://lccn.loc.gov/2016017082

Published by Nova Science Publishers, Inc. † New York

To my wife, children, and parents

In recognition of their sacrifices, love, patience and affection that are beyond description

CONTENTS

PREFACE

As compared to conventional DC and synchronous motors, induction motors (IMs) have been widely used in industry for variable speed drives due to some of the advantages; these include robust construction, reliability, high power to weight ratio, and high torque to inertia ratio. However, the IM drives are nonlinear, time-varying, and complex systems due to their characteristics. Since 1980, the FOC method has been applied for high performance control of the IM drive. Due to some important characteristics of variable-structure control (VSC) (or in particular, SMC system such as robustness to system parameter and fast dynamic response), it is applied in IM drives.

This book develops sliding-mode and fuzzy logic based speed controllers for an indirect field-oriented control (IFOC) of an induction motor (IM) drive. The structures of these speed controllers are based on the boundary layer approaches, which can be referred to as boundary layer fuzzy controller (BLFC) and fuzzy sliding-mode controller (FSMC). Likewise, they may be combined to implement a high performance IM drive. Essentially, the boundary layer approach is the most popular method to reduce the chattering phenomena, which leads to trade-off between control performances and chattering elimination for uncertain nonlinear systems. Induction motor drives face the possibility of large uncertainty, such as external load disturbance and some variations of critical parameters in their working conditions. Therefore, these intelligent speed controllers are designed so that they can eliminate chattering completely while optimum tracking performance is achieved for the IM drive.

The main objective of this book is to provide the necessary background to work with fuzzy sliding mode speed controller to implement high performance drive. Once the material in this book has been mastered, the reader will be able to apply a fuzzy sliding mode speed controller to his or her problems for high performance drive.

To facilitate this goal, the chapter-1 introduce about IM and its control techniques as well as problem identification and book objectives have been included in this chapter.

The Chapter-2 provides a literature review and summary of the current research in IM drive incorporating different control techniques such as, intelligent and nonlinear speed controllers are described in chapter-2. The IFOC of IM drive is also reviewed in detail in chapter 2. Moreover, motivation of the current work is presented. Finally, the theoretical bases that can be used in order to complete a book project are explained in the last section.

The Chapter-3 presents a novel fuzzy sliding-mode control for chattering-free and robust induction motor drive. Designing the FSMC system and the proposed speed controller are also shown. Moreover, the simulation results are shown to validate the proposed sliding-mode speed controller. The concluding remarks are finally made in final Section.

In the fourth Chapter, the proposed NBLFC control based on the boundary layer approach for an indirect field-oriented control (IFOC) of an induction motor (IM) drive is explored. The real-time implementation of the conventional PI, the BLFC, and the proposed NBLFC based speed controller of IM drives using DSP controller board ezdspF28335 for a laboratory 1 kW motor is presented. Extensive computer simulations and numerous experimental results for the conventional PI, BLFC, and the proposed NBLFC controllers are shown in this chapter. A real-time performance comparison of the proposed NBLFC with the conventional PI and BLFC controller is also presented in this chapter.

The Chapter-5 presents how to teach and simulate an adjustable speed drive of IM using Simulink blocks for an indirect field-oriented control (IFOC) algorithm. The optimized Simulink blocks is used to design the precise Simulink models so that tracking of the speed and flux in the presence of an uncertain load disturbance and parameter variations is provided. The effectiveness of the adjustable IM drive is verified by simulation the tuned PI controller results at different operating conditions over a wide speed range.

Chapter-6 highlights the required hardware and software for real-time implementation of high performance IM drive. The detail real-time implementation of high performance induction motor drive using digital signal processor board TI TMS320F28335 is also presented.

The summary of the contribution and the concluding remarks of this work are made in chapter 7. In addition, the originality of the work is explored in the next section. Furthermore, some suggestions for the future works and studies are explored in this chapter. Finally, all related Appendices are listed at the end of this book.

ACKNOWLEDGMENTS

First and foremost, all praise is for Allah, the Almighty, the creator of the heavens and the earth and may his peace and blessings be on the last prophet Muhammad, on his family and companions.

I would like to take this opportunity to express my deepest appreciation and gratitude to Professor M. Nasir Uddin from Lakehead University in Canada for valuable discussions and useful suggestions and help for completion of this book.

I would like to appreciate my parents and family to whom my dedication is due to them, especially, my mother, who encouraged me all the time to continue education. Also, I would like to remember my father who passed away and he did his best for me to reach this point of my life. I would like to appreciate my wife, for her encouragement and support, without which it would not have been possible for me to complete this work.

With best wishes to all of them

Ali Saghafinia*, PhD
Author,
Assistant Professor
Electrical Engineering Department,
Majlesi Branch,
Islamic Azad University,
Esfahan, 8631656451 Iran
Email: saghafi_ali@yahoo.com

* Corresponding Author: Email: saghafi_ali@yahoo.com.

LIST OF ABBREVIATIONS

v_a, v_b, v_c	Three-phase voltages
$v^e{}_{ds}, v^e{}_{qs}$	Synchronous frame stator voltages
$v^e{}_{dr}, v^e{}_{qr}$	Synchronous frame rotor voltages
i_a, i_b, i_c	Three-phase currents
$i^e{}_{ds}, i^e{}_{qs}$	Synchronous frame stator currents
$i^e{}_{dr}, i^e{}_{qr}$	Synchronous frame rotor currents
$\varphi^e{}_{ds}, \varphi^e{}_{qs}$	Synchronous frame stator fluxes
p	Differential operator
$\dot{x}$	First derivative of x
$\ddot{x}$	Second derivative of x
$\overset{*}{x}$	Command value of x
$\lvert x \rvert$	Absolute value of x
ω_e	Synchronous speed
ω_{r}	Rotor speed
ω_{sl}	Angular slip velocity
P	Poles number of motor
R_s, R_r	Stator and rotor resistance, respectively
L_s, L_r	Stator and rotor inductance, respectively
L_m	Mutual inductance

σ	Total leakage factor $\sigma = 1 - \frac{L_m^2}{L_s L_r}$
e	Error between actual and command speed
Δe	Change of speed error
θ_r	Rotor position
s	Laplace operator
$S(t)$	Time-varying surface of sliding mode
h	Positive constant for sliding surface
$u(t)$	Control effort
$u_{eq}(t)$	Equivalent control
$u_r(t)$	Reaching control
$L(t)$	Lumped uncertainty
$k(t)$	Control gain
τ	Integral constant time of system
$\psi(t)$	Boundary layer thickness
$V(t)$	Lyapunov function
η	Positive constant of stability threshold
t_{reach}	Reaching phase
T_e	Motor torque
T_L	Load torque
J_r	Rotor inertia
B_r	Friction factor
$\Delta S(t)$	Alteration of slide surface
λ_k	Coefficient of adaptive law for AFSMC
$\hat{k}$	Estimation of gain control
$k_d(t)$	Desired control gain
υ	Cut-off filter frequency
KPd, KPq	Proportional term of PI controller
KId, KIq	Integral term of PI controller
KCd, KCq	Correction term of PI controller
V	Lyapunov function

LIST OF ACRONYMS

ADC	Analog to Digital Convertor
AFSMC	Adaptive Fuzzy Sliding Controller
ASIC	Application Specific Integrated Circuit
CISC	Complex Instruction Set Computing
COA	Center of Area
DMC	Digital Motor Control
DSC	Digital Signal Controller
DSP	Digital Signal Processor
ES	Embedded System
FLC	Fuzzy Logic Controller
FOC	Field-Orientation Control
IFOC	Indirect Field- Orientation Control
IM	Induction Motor
IPMSM	Interior Permanent Magnet Synchronous Motor
LMA	Loss- minimization algorithm
MF	Membership function
N	Negative
NB	Negative Big
NBLFC	Novel Boundary Layer Fuzzy Controller
NM	Negative Medium
NS	Negative Small
P	Positive
PB	Positive Big
PDF	Pseudo- Derivative-Feedback
PI	Proportional Integral
PID	Proportional Integral Derivative

PM	Positive Medium
PMSM	Permanent Magnet Synchronous Motor
PS	Positive Small
PWM	Pulse Width Modulation
Rpm	Revolution Per Minute
RISC	Reduced Instruction Set Computing
Sat	Saturation Function
SMC	Sliding-Mode Control
Sgn	Sign Function
SVM	Space Vector Modulation
VBLSMC	Variable Boundary Layer SMC Controller
VHDL	Very High Speed Circuit Description Language
VLSI	Very Large Scale Integration
VSC	Variable Structure Control
VSI	Voltage Source Inverter
Z	Zero
ZE	Zero

LIST OF TABLES

LIST OF FIGURES

Chapter 1

INTRODUCTION

ABSTRACT

This chapter presents an introduction about induction motor (IM) and its control techniques as well as problem identification and some nessary information for apllying the IM drive and book objectives have been included in this chapter.

Keywords: induction motor drive, vector control, applied inverter, speed control, control techniques

1.1. PREAMBLE

The electrical motors convert more than 50% of electrical energy into mechanical energy. Over the years, induction motors (IM's) have been widely used in industry for variable speed drives due to some of the advantages such as robust construction, low cost, low maintenance, and reliability [1].

The DC motor was first invented in 1873. The DC motor decouples torque and flux naturally which had been applied extensively for high performance variable electrical drives. However, some negative characteristics of DC machines such as large dimension, heavy weight to power ratio, high cost of brushes and commutators, frequent maintenance requirement and limited range of operation introduce them as an expensive solution. Moreover, undesired sparking which is made by the commutator-brush-assembly reduces the drive efficiency and also limits the application of DC motors [2].

Due to these inherent drawbacks of DC motors, the researchers developed squirrel cage IMs, which overcome these problems.

1.2. Induction Motor

Among the two types of IMs, squirrel-cage and wound rotor, the squirrel-cage IMs are used as a workhorse in the industry from fractional horsepower (HP) to multi-megawatt capacity because of some advantages such as economical, rugged, and reliable. However, the IM drives are nonlinear, time-varying and complex systems due to their characteristics [3-5] which require the complex control, circuitry, and inverter over sizing [6].

Based on torque-speed characteristic curve, the squirrel-cage IMs are classified into five groups such as, A, B, C, D, and E. Rotor resistance plays an important role in this classification. Due to the low rotor resistance of IMs in class A, IMs need high starting current and low starting torque, which leads to high efficiency of this motor in low slip frequency. Class B machines with high leakage inductance rotor have lower starting current and torque than class A machines with regard to better slip frequency characteristic. Therefore, they are applied to constant-speed industrial drives. Class C and D machines are designed so that their high rotor resistance leads to the low starting current and high starting torque in comparison with the classes A & B. The last class E machines include high efficiency machines to high performance IM drives.

1.3. Induction Motor Drives

Electric motor drives are applied widely in industrial applications such as pumps, fan, paper and textile mills, elevators, electric vehicle and subway transportation, home applications, servos and robotics, ship propulsion, etc. Some characteristics are required to design and implement a high performance electrical drive such as general knowledge about machine performance, the dynamic model, and parameter variations. Nowadays, high performance electric motor drives are an essential requirement for new industrial applications. The high performance electric motor drives have some characteristics such as: high reliability, high energy transformation efficiency, and quick response of torque, position and speed, robust to parameter variations and external load disturbance, low weight and less expensive. A general structure of electric motor drives has shown in Figure 1.1.

The variable frequency IM drives can be mainly classified into categories namely scalar control and vector control. Classification of these methods is shown in Figure 1.2. Each control technique could be open loop (no feedback) or closed loop (with feedback). The IM control can further be classified as V/f (constant flux) and flux weakening (torque control). The V/f method is used to control the motor up to rated speed. In order to control the motor above rated speed, the flux must be weakened so that the motor operates within rated voltage and current capacities. The V/f method, which is the most common scalar method, maintains the stator flux at a constant level by increasing the voltage and frequency of stator simultaneously so that the rate of voltage/frequency remains constant. This method operates correctly during steady-state operation.

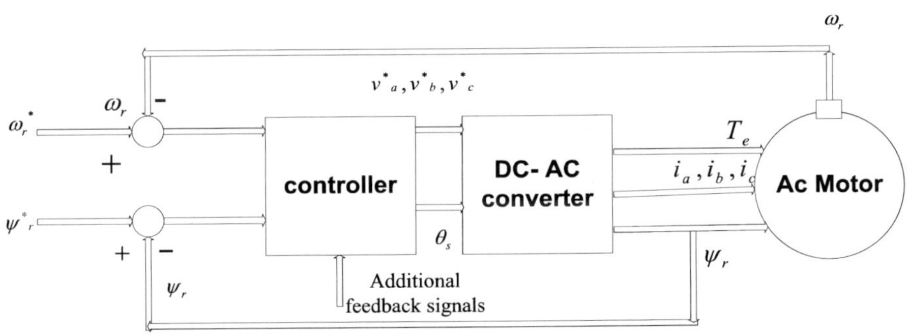

Figure 1.1. The general block diagram of electric motor drive.

The field orientation control decouples the flux and torque controls so that the central task becomes easier in both steady and transient states. Thus, the IM behaves like a separately excited DC motor while maintaining its general advantages over DC motor. Therefore, the vector control method is preferred to the scalar method, which will be also be used in this research.

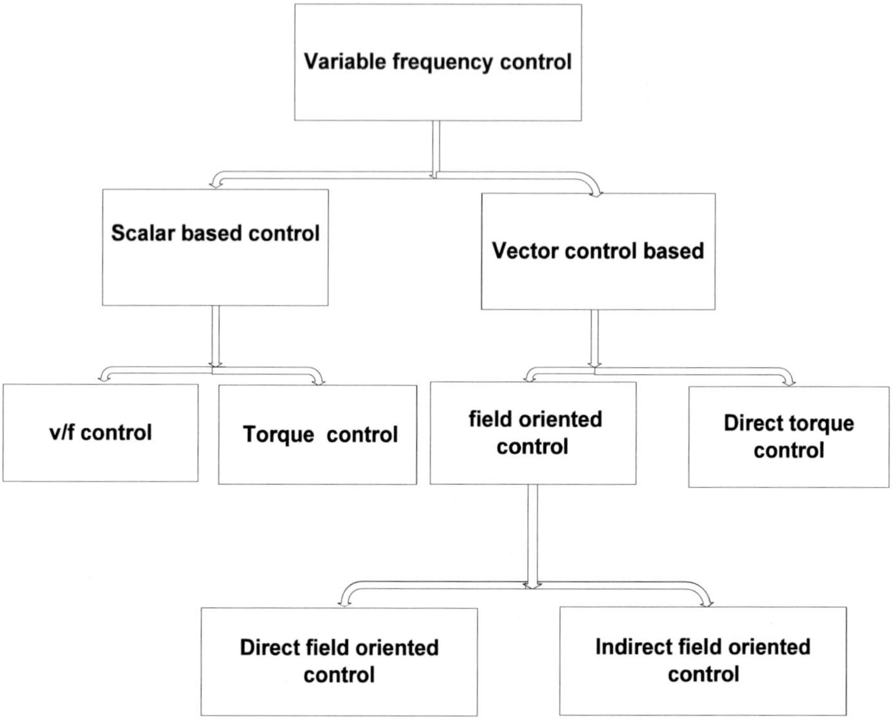

Figure 1.2. The classification of control method of IM drives.

1.3.1. Vector Control of IM Drives

The flux vector is required to implement the field-oriented control, in the two general methods of the vector control or field-oriented control. The mechanisms can differ, however, depending on the acquirement method of the flux vector. The direct method or flux feedback method is one of them and another one is called the indirect method. Due to the use of sensors to detect the air gap flux in the direct method, this method requires additional electronics, extra wiring and space, frequent maintenance and careful mounting, which causes decrease in the inherent robustness and reliability of the drive and also increased cost. A major drawback with the direct orientation method is the difficulty it poses to measure the air gap flux at very low speed. That is why, the development of the alternative indirect method becomes an important research topic continually [7-10].

The indirect field oriented control as an alternative method to measure the flux position without using any flux sensor, and utilizes the rotor speed and the angular slip frequency to make the unit vector signals for achieving the flux orientation. In fact, the motor current components are decoupled by estimation of the slip speed, which requires a suitable knowledge of the rotor time constant. The accuracy of this method depends on the precise estimation of the motor parameters and rotor time constant [11, 12]. However, changes in these parameters often cause field orientation detuning and degrade the electrical drive performance. Thus, the torque capability and efficiency of the drive are reduced in steady state. Also, the torque/ampere capability is significantly decreased due to the inverter current limits and causes unsatisfactory drive performance especially to fast dynamic speed command [13, 14]. Moreover, disturbances such as external load torque and unmodeled dynamics have effect on the drive performance [15]. In order to achieve indirect vector control of high performance from IM drive sophisticated speed control method is required [16].

1.3.2. The Applied Inverters for IM Drives

In general, inverters receive the DC voltage as input from the rectifier circuits and convert it to an AC variable voltage with fixed or variable frequency depending on their applications. The power semiconductor switches used for inverter are GTOs, BJTs, IGBTs, Power MOSFETs, IGCTs, etc. Depending on the existence of voltage or current source as source for DC link, there are two models of the inverter, which are known as voltage source inverter (VSI) and current source inverter (CSI) respectively. The VSI inverters are the best option to speed control IM drives because of some attractive feature as follows [1]: (1) existence of a large leakage inductance arising from motor filter harmonics in the VSI inverter while the current transfer interval limits the fundamental frequency of operation; (2) despite the symmetric switching device in the CSI inverter unlike the VSI inverter, the commutation time is affected by the external circuit and consequently the switching rate can be reduced; (3) the dynamic response of VSI inverters is better than the CSI inverters; (4) to implement the CSI inverters unlike the VSI inverters, it requires a minimum load which always remains to be connected.

Voltage–fed converters have extensive applications such as AC motor drives, AC uninterruptible power supplies (UPS), active harmonic filter, and induction heating. Fast and energy-efficient control of the electrical drives are leaded the increasing apply of power electronic converters. Voltage source inverters (VSI) are applied in field oriented control (FOC) of three-phase squirrel cage IM drives to regulate the speed by changing the frequency and voltage [17, 18].

Generally, Six-step method and pulse –width modulation (PWM) techniques are two popular methods for FOC of IM drives by using the VSI inverters. Three phase six-step inverters as usual inverters have some advantages such as, simplistic, the low switching loss because of six switching per cycle of fundamental frequency, elimination of third order and its multiple harmonic from current wave. However, the existing lower order harmonics of the six-step voltage wave, which causes large distortions of the current wave and voltage control by the line-side rectifier, are limitations of this switching method. The VSI inverters along with the PWM techniques with regard to parameters such as cost, efficiency, and transient state are the best option for application in high performance IM drives [1].

The PWM techniques are able to control output voltage as well as optimize the harmonics due to the power electronic switching. Among PWM techniques including sinusoidal PWM (SPWM), selected harmonic elimination (SHE) PWM, minimum ripple current (MRC) PWM, space vector PWM (SVM), random PWM (RPWM), hysteresis bound current control PWM, and sinusoidal PWM with instantaneous current control, the most popular method of them to IM drives is the SVM [1]. The SVM method considers interaction of the phases and optimizes the harmonic content of current signal while the amplitude of the output voltage is the highest in comparison with other methods and with same DC link voltage.

1.3.3. Implementation Techniques for IM Drives

Traditionally, the IM control was designed based on analog components because of some advantages such as, simple design and low cost. Due to aging components and temperature existence, reliability of the system is reduced. Moreover, analog components increase tolerance issues and their upgrades are very difficult so that the implemented system performance is very low [19]. Digital systems are able to improve analog designs and system upgrade by using software due to their advantages such as increasing of calculation speed and precise control, etc. They assist IMs to be benefitted from an enhanced real-time algorithm, thus leading to high performance. To implement the complex electrical drives by using digital systems, the advanced microcontrollers, digital signal processors (DSPs), and field programmable gate array (FPGA) are applied for many applications [20-24].

Due to the new design of control strategies and the technological progress of semiconductor, the suitable selection of the digital systems can increase the system performance. Among the digital systems, the DSPs are better options for electrical drive applications with some of their advantages as follows: (1) algorithms written in high level languages can be easily implemented; (2) it produces a high resolution PWM outputs to control switching inverters; (3) decreasing the number of look-up tables in spite of the intelligent controllers which reduces the required memory; (4) to eliminate narrow-band mechanical resonance, which required sharp-cut-off notch filters.

1.4. Speed Control Structure of IM Drives

Generally, in the domain of drives control, the methods of control can be classified into three main categories such as, fixed gain methods or linear methods, adaptive, artificial intelligence methods and combination of them may also be used depending on their applications. Linear controllers which include the proportional integral controller (PI), proportional-integral-derivative (PID), and pseudo- derivative-feedback (PDF) are used as the most common approaches in industrial applications because of some advantages such as simple structure, simplicity, and many advanced algorithms and strategies such as model predictive control can be applied to this controller [25, 26]. However, the IM drives are nonlinear, time variant, complex, and uncertain systems. The linear controller like PI controller is unable to deal with any certainties. In fact in case of the indirect FOC induction motor drive, the PI controller is sensitive to the system parameter variations, insufficient to reject external load disturbances, and not robust to inertia increase and rotor resistance variations [27, 28].

Several nonlinear adaptive methods are able to adapt or upgrade the PI controllers such as model reference adaptive control (MRAC) [29], variable structure control (VSC) [30], and self-tuning PI controllers [31], etc. To implement the above-mentioned speed controller methods, it is required to understand the exact mathematical model.

Due to some important characteristics of variable-structure control (VSC) or in particular, sliding –mode control (SMC) system such as robustness to system parameters and fast dynamic response, it is applied in IM drives, [32, 33]. However, the SMC based system suffers from a chattering problem in steady state. The chattering makes various undesirable effects such as current harmonics and torque pulsation. To reduce or eliminate the chattering some methods have already been proposed by the researchers [34-37]. Sometimes, these controllers are used along with the artificial intelligent controllers such as fuzzy sliding-mode controller (FSMC), boundary layer fuzzy controller (BLFC), artificial neural network controller, and neural-fuzzy controller to reduce the influence of the mathematical model [38-40]. The mentioned fuzzy controller on the basis of the SMC system have some advantages such as, decreasing the number of fuzzy rules and relaxation of the uncertainty bound requirement [41, 42]. These speed controllers were designed based on fixed and variable boundary layer SMC, which named fuzzy sliding-mode (FSMC), fixed boundary layer SMC (FBLSMC) (or non-fuzzy SMC), and boundary layer fuzzy (BLFC) controllers [35, 43-45].

1.5. Problem Statement

According to the previous section, the SMC based adaptive methods are suitable for speed control of the high performance IM drive. However, the conventional SMC suffers from chattering in steady state. Due to the advantages of the fuzzy controllers on the basis of the SMC system, the fuzzy controllers were used to improve the chattering and the dynamic performance of the SMCs based speed controller drives [35, 43-45]. However, these works cannot eliminate the chattering in torque and speed responses of the IM drive at different operation conditions and uncertainties due to the lack of proper development of those controllers. In addition, despite the many advantages of IM, it is highly nonlinear due to magnetic saturation and armature reaction effects. Furthermore, the IM always runs below synchronous speed in the motor mode. Thus, the control of IM becomes complicated. On the other hand, the real-time high performance IM drive must provide fast and accurate speed responses, fast disturbance rejection, and insensitivity to parameter variations. Therefore, careful consideration must be done to develop control algorithms for real-time high performance drives to overcome these problems. Hence, the applications of IMs for variable speed drives remain an engineering challenge.

1.6. Objectives of the Book

In this book, the IFOC method is used to develop high performance IM drive. The high performance drives require fast and accurate speed responses, fast disturbance rejection, and insensitivity to parameter variations. In order to achieve high performance drive, hybrid intelligent control techniques involving fuzzy logic and SMC will be developed. Particular objectives of this book are as follows:

- Development of the FLC and SMC based hybrid speed controllers such as novel fuzzy sliding mode (NFSMC) and novel boundary layer sliding-mode (NBLFC) controllers for an IM drive system so that the high dynamic speed response can be achieved without chattering issue.
- Investigation of the proposed NFSMC and NBLFC based IM drives using Matlab/Simulink.
- Real-time implementations and performance investigations of the conventional SMC and the proposed NBLFC based IM drive using the Texas Instrument' DSP TMS320F28335'.

REFERENCES

[1] B. K. Bose, *Modern power electronics and AC drives*: Prentice Hall PTR USA, 2002.

[2] R. M. Crowder, *Electric drives and their control*: Clarendon Press, 1998.

[3] L. Qaseer, S. Purushothaman, and F. de Leon, "Closed-Form Analysis of Squirrel-Cage Induction Motors with Anisotropic Modeling of Stator and Rotor," *Energy Conversion, IEEE Transactions on,* vol. PP, pp. 1-8, 2012.

[4] M. Ojaghi, J. Faiz, M. Kazemi, and M. Rezaei, "Performance Analysis of Saturated Induction Motors by Virtual Tests," *Education, IEEE Transactions on,* vol. PP, pp. 1-1, 2011.

[5] A. Saghafinia, S. Kahourzade, A. Mahmoudi, W. Hew, and M. N. Uddin, "Broken rotor bar fault detection of 3-phase induction motor using online adaptive continuous wavelet transform and fuzzy logic," *International Review of Electrical Engineering-IREE,* vol. 7, pp. 4383-4394, 2012.

[6] U. Bakshi and V. Bakshi, *Electrical Circuits and Machines*: Technical Publications, 2009.

[7] M. S. Zaky, M. Khater, H. Yasin, and S. S. Shokralla, "Speed-Sensorless Control of Induction Motor Drives (Review Paper)."

[8] J. Holtz, "Sensorless control of induction motor drives," *Proceedings of the IEEE,* vol. 90, pp. 1359-1394, 2002.

[9] A. Saghafinia and H. W. Ping, "High performance induction motor drive using fuzzy self-tuning hybrid fuzzy controller," in *Power and Energy (PECon), 2010 IEEE International Conference on*, 2010, pp. 468-473.

[10] A. Saghafinia, H. W. Ping, and M. Rahman, "High performance induction motor drive using hybrid fuzzy-pi and pi controllers: A review ",*International Review of Electrical Engineering-Iree,* vol. 5, pp. 2000-2012, 2010.

[11] M. Moallem, B. Mirzaeian, O. A. Mohammed, and C. Lucas, "Multi-objective genetic-fuzzy optimal design of PI controller in the indirect field oriented control of an induction motor," *Ieee Transactions on Magnetics,* vol. 37, pp. 3608-3612, 2001.

[12] Z. Zhen-Yu, M. Tomizuka, and S. Isaka, "Fuzzy gain scheduling of PID controllers," *Systems, Man and Cybernetics, IEEE Transactions on,* vol. 23, pp. 1392-1398, 1993.

[13] K. B .Nordin, D. W. Novotny, and D. S. Zinger, "The influence of motor parameter deviations in feedforward field orientation drive systems," *IEEE Transactions on Industry Applications,* pp. 1009-1015, 1985.

[14] L. Zhen and L. Xu, "On-line fuzzy tuning of indirect field-oriented induction machinedrives," *IEEE Transactions on Power Electronics,* vol. 13, pp. 134-141, 1998.

[15] M. Masiala, B. Vafakhah, J. Salmon, and A. M. Knight, "Fuzzy Self-Tuning Speed Control of an Indirect Field-Oriented Control Induction Motor Drive," in *41st Annual Meeting of the IEEE-Industry-Applications-Society*, Tampa, FL, 2006, pp. 1732-1740.

[16] A. Consoli, G. Scarcella, and A. Testa, "Slip-frequency detection for indirect field-oriented control drives," *IEEE Transactions on Industry Applications,* vol. 40, pp. 194-201, 2004.

[17] A. Abdelsalam, M. Masoud, S. Finney, and B. Williams, "Vector control PWM-VSI induction motor drive with a long motor feeder: performance analysis of line filter networks," *IET Electric Power Applications*, vol. 5, p. 443, 2011.

[18] D. Morinigo-Sotelo, L. Garcia-Escudero, O. Duque-Perez, and M. Perez-Alonso, "Practical aspects of mixed-eccentricity detection in PWM voltage-source-inverter-fed induction motors," *Industrial Electronics, IEEE Transactions on*, vol. 57, pp. 252-262, 2010.

[19] T. Instruments, "Digital Signal Processing Solution for AC Induction Motor," *Application Note Literature Number BPRA043,* 1996.

[20] H. Li, "A stochastic-based FPGA controller for an induction motor drive with integrated neural network algorithms," *Industrial Electronics, IEEE Transactions on,* vol. 55, pp. 551-561, 2008.

[21] C. T. Kowalski and J. D. Lis, "Speed sensorless DTC control of the induction motor using FPGA implementation," *COMPEL: The International Journal for Computation and Mathematics in Electrical and Electronic Engineering,* vol. 29, pp. 109-125, 2010.

[22] Y. T. Kao and C. H. Liu, "Analysis and design of microprocessor-based vector-controlled induction motor drives," *Industrial Electronics, IEEE Transactions on,* vol. 39, pp. 46-54, 1992.

[23] F. J. T. E. Ferreira, "Strategies to improve the performance of three-phase induction motor driven systems," 2009.

[24] S. Y. Wang, C. L. Tseng, C. Y. Chang, C. C. Yeh, and J. H. Chou, "Adaptive proportional-integral speed controller for direct torque control of induction motor using neural networks," 2010, pp. 145-147.

[25] S. Tzafestas and N. P. Papanikolopoulos, "Incremental fuzzy expert PID control," *Ieee Transactions on Industrial Electronics,* vol. 37, pp. 365-371.1990 ,

[26] Z. Y. Zhao, M. Tomizuka, and S. Isaka, "Fuzzy gain scheduling of PID controllers," *IEEE Transactions on Systems, Man and Cybernetics,* vol. 23, pp. 1392-1398, 1993.

[27] M. Cheng, Q. Sun, and E. Zhou, "New self-tuning fuzzy PI control of a novel doubly salient permanent-magnet motor drive," *IEEE Transactions on Industrial Electronics,* vol. 53, pp. 814-821, 2006.

[28] F. Barrero, A. Gonzalez, A. Torralba, E. Galvan, and L. G. Franquelo, "Speed control of induction motors using a novel fuzzy sliding-modestructure," *IEEE Transactions on Fuzzy Systems,* vol. 10, pp. 375-383, 2002.

[29] S. Maiti, C. Chakraborty, Y. Hori, and M. C. Ta, "Model reference adaptive controller-based rotor resistance and speed estimation techniques for vector controlled induction motor drive utilizing reactive power," *Industrial Electronics, IEEE Transactions on,* vol. 55, pp. 594-601, 2008.

[30] G. Abad, M. A. Rodríguez, and J. Poza, "Two-level VSC based predictive direct torque control of the doubly fed induction machine with reduced torque and flux ripples at low constant switching frequency," *Power Electronics, IEEE Transactions on,* vol. 23, pp. 1050-1061, 2008.

[31] X. H. Kong, B. J. Zhang, X. H. Mao, Y. F. Chen, and C. Y. Song, "Design and Application of Self-Tuning PI Controller," *Applied Mechanics and Materials,* vol. 43, pp. 160-164, 2011.

[32] V. I. Utkin, "Sliding mode control design principles and applications to electric drives," *Industrial Electronics, IEEE Transactions on,* vol. 40, pp. 23-36, 1993.

[33] A. Bartoszewicz and A. Nowacka-Leverton, "SMC without the reaching phase-the switching plane design for the third-order system," *Control Theory & Applications, IET,* vol. 1, pp. 1461-1470, 2007.

[34] F. Barrero, A. Gonzalez, A. Torralba, E. Galvan, and L. G. Franquelo, "Speed control of induction motors using a novel fuzzy sliding-mode structure," *IEEE Transactions on Fuzzy Systems,* vol. 10, pp. 375-383, 2002.

[35] R. Shahnazi, H. M. Shanechi, and N. Pariz, "Position control of induction and DC servomotors: a novel adaptive fuzzy PI sliding mode control," *Energy Conversion, IEEE Transactions on,* vol. 23, pp. 138-147, 2008.

[36] Z. Jinhui, S. Peng, and X. Yuanqing, "Robust Adaptive Sliding-Mode Control for Fuzzy Systems with Mismatched Uncertainties," *Fuzzy Systems, IEEE Transactions on,* vol. 18, pp. 700-711, 2010.

[37] Z. H. Salih, K. S. Gaeid, and A. Saghafinia, "Sliding Mode Control of Induction Motor with Vector Control in Field Weakening," *Modern Applied Science,* vol. 9, p. p276, 2015.

[38] M. A. Fnaiech, F .Betin, G. A. Capolino, and F. Fnaiech, "Fuzzy logic and sliding-mode controls applied to six-phase induction machine with open phases," *Industrial Electronics, IEEE Transactions on,* vol. 57, pp. 354-364, 2010.

[39] A. Kusagur, S. Kodad, and B. V. S. Ram, "Modeling, Design & Simulation of an Adaptive Neuro-Fuzzy Inference System (ANFIS) for Speed Control of Induction Motor," *International Journal of Computer Applications IJCA,* vol. 6, pp. 29-44, 2010.

[40] Y. K. Kim and G. J. Jeon, "Error reduction of sliding mode control using sigmoid-type nonlinear interpolation in the boundary layer," *International Journal of Control, and Systems,* vol. 2, pp. 523-529, 2004.

[41] N. Yagiz, Y. Hacioglu, and Y. Taskin, "Fuzzy sliding-mode control of active suspensions," *Industrial Electronics, IEEE Transactions on,* vol. 55, pp. 3883-3890, 2008.

[42] C. L. Kuo, T. H. S. Li, and N. R. Guo, "Design of a novel fuzzy sliding-mode control for magnetic ball levitation system," *Journal of Intelligent and Robotic Systems,* vol. 42 ,pp. 295-316, 2005.

[43] T. Kuo-Yang, L. Tsu-Tian, and W. Chi-Hsu, "Design of a New Fuzzy Suction Controller Using Fuzzy Modeling for Nonlinear Boundary Layer," *Fuzzy Systems, IEEE Transactions on,* vol. 13, pp. 605-616, 2005.

[44] T. Orowska-Kowalska, M .Kaminski, and K. Szabat, "Implementation of a Sliding-Mode Controller with an Integral Function and Fuzzy Gain Value for the Electrical Drive with an Elastic Joint," *Industrial Electronics, IEEE Transactions on,* vol. 57, pp. 1309-1317, 2010.

[45] L. Viet Quoc, C. Han Ho, and J. Jin-Woo, "Fuzzy Sliding Mode Speed Controller for PM Synchronous Motors with a Load Torque Observer," *Power Electronics, IEEE Transactions on,* vol. 27, pp. 1530-1539, 2012.

Chapter 2

LITERATURE REVIEW

ABSTRACT

This chapter presents a comprehensive review of the control techniques for high performance IM drives. Moreover, the different implementation techniques for IM drives are also reviewed. Morover, the literature related to adaptive, intelligent, and nonlinear speed controller for high performance IM drives is explored. Finally, the theoretical bases including mathematical model of IM for IFOC, induction motor equation, Indirect field orinted control, the principle of fuzzy logic, sliding mode speed control, and conventional SMC system are included in the last section.

Keywords: literature review, direct vector control, indirect vector control, fuzzy sliding mode control, boundary layer fuzzy controller, conventional sliding-mode speed control

2.1. PREAMBLE

The induction motor was invented in 1888 by Nikola Tesla [1]. Dolivo Dobrowolski extended its structure in 1890 which is known as squirrel-cage motor [2]. Although unchanged in general appearance during these times until now, the design of these machines has been improved. Thus, its variety characteristics lead to increase the squirrel-cage motor applications in modern industries. High performance electrical motor drives are considered as an essential requirement for modern industrial applications [3]. In the past decades, DC motors have been extensively applied for this purpose. However, their disadvantages prevent implementation high performance drives, which were described in Section 1.2. Due to the advantages of IM as compared to DC motors and synchronous motors such as robust construction, low cost, low maintenance, and reliability to inertia ratio, numerous attempts have been made to apply induction motor instead DC motors in modern drive applications [4, 5]. Researchers have paid attention to developing controllers for IM drives due to its wide use in industry [6]. In recent years, the progress in semiconductor switching devices assists the researchers to utilize closed-loop vector control technique for IM drives.

2.2. The Control Approaches of the IM Drive

Essentially, the electric motor drives are controlled by two general approaches, namely the scalar and the vector control method. In early days researchers controlled the IM using a scalar method like volt-hertz control [7, 8]. In this control, the magnitude and frequency of the stator voltages are calculated and they are controlled along with current and flux vector under the steady-state condition, which results in poor dynamic performance. Unlike the scalar method, the FOC method controls the magnitude and frequency of voltage, current, and flux vector to steady state as well as transient state. The stator current components obtained from the FOC method are oriented in flux component and the torque component. The flux vector can be either stator flux, air gap or mutual flux, or rotor flux. The orientation of the stator current with regard to the stator, rotor, and air gap flux is examined and the relative advantages and its developments are reported [9, 10]. The salient advantages of the FOC method are simplification of flux computation, implementation of field weakening operation, fast torque response, and the decoupling of torque and flux characteristics [11]. Thus, the IM can be controlled like a separately excited DC motor. Moreover, with the recent advances of power semiconductor technologies and various intelligent control algorithms, an effective control method based on the FOC can be implemented completely in real-time industrial drive applications. So, the vector control approach is the suitable method and better option than the scalar control method for high performance IM drives [12].

Owing to the flux acquisition method, there are two approaches to implement FOC method precisely. The first one is known as the direct method

[4] and other one is known as the indirect method [13]. The universal FOC was developed by Doncker and Novotny [14] so that it is applicable to both methods and the generalized approach is used for indirect field oriented control of induction and synchronous motors [15]. The two mentioned methods are reviewed and discussed in the next sub-sections.

2.2.1. The Direct Vector Method of the IM Drive

In the direct FOC method or flux feedback method, The air gap flux is directly measured by using some sensors such as Hall probes and search coil or trapped stator winding [16, 17] or using observed based variables such as, stator voltage, current, and speed [18, 19]. Due to the use of a sensor to sense the air gap flux in the direct method, it requires additional electronics, extra wiring and space, frequent maintenance and careful mounting which causes decrease in inherent robustness and reliability of the drive and also increase in cost. That is why, the development of the alternative indirect method becomes an important research topic continually [20, 21]. Several of the flux observers are applied to estimate and improve the flux response with less sensitivity to motor parameters [22]. However, a major drawback to the direct FOC method is the necessary to measure the air gap flux at very low speed. Some amount of research has been done to overcome this problem by observing stator flux and measuring rotor position [23].

2.2.2. The Indirect Vector Control Method of IM Drive

The IFOC method as an alternative method to measure the flux position without sensor, applies the rotor speed and the slip angular frequency to construct the unit vector signals for achieving the flux orientation. In fact, the motor current components are decoupled by estimating the slip speed, which requires a suitable knowledge of the rotor time constant. The accuracy of this method depends on the precision of the motor parameters and rotor time constant estimation [24]. However, changes in these parameters, often cause field orientation detuning and degrade the electrical drive performance. In this state, the torque capability and efficiency of the drive are reduced in steady state. Also, the torque/ampere capability is significantly decreased due to the inverter current limits and cause unsatisfactory drive performance especially, to fast changing speed command [25]. Moreover, the disturbances such as external load torque and unmodeled dynamics affect the drive performance [26]. However, the suitable design of speed controller can overcome the mentioned disadvantages [6].

2.3. The Applied VSI Inverters for IM Drives

In high performance IM drives, the suitable choosing of the VSI inverters is very important [3]. Elimination of harmonics from the output voltage is the most important issue in a multi-level inverter. One of the most important characteristics of inverters is total harmonic distortion (THD) of the current or voltage. The output of the inverter must meet maximum THD boundaries as specified by IEEE [27]. The existence of lower order harmonics in six-step voltage wave for the three-phase inverter lead to apply the PWM techniques for FOC of VSI fed IM drive. Besides many topologies such as neutral point clamped, flying capacitor, H-bridge, cascaded with separated DC source, several control and modulation strategies such as SPWM, SHE PWM, SVM, etc. have been applied to modern industrial drive applications.

In 1964, Schonung and Stemmler proposed PWM approach which is made by using sine triangle intersection [28]. The sinusoidal PWM technique is widely used in the industry due to its simplicity in implementation. The main drawback of this technique is inefficient use of the DC bus voltage (maximum 78.5% of the DC bus voltage). For efficient use of DC bus voltage, the space vector modulation (SVM) technique is introduced by Pfaff, et al. in 1984 [29] and improved by Vander, et al. in 1988 [30]. Due to the development of microcontrollers, this technique can also be easily applied in modern industrial drive applications. Among the above mentioned PWM techniques, the SVM method is the best one because it considers interaction of the phases and optimizes the harmonic content of current signal, and switching losses while the amplitude of the output voltage is the highest in comparison with the other methods by using the same quantity of DC link voltage [31, 32]. In recent year, the SVM technique has been applied for VSI fed high performance IM drives [33, 34]. Therefore, the SVM technique is applied in this book.

2.4. Implementation of IFOC Based IM Using Microcontroller/DSP

Due to the recent advances in power semiconductor technology, microcontrollers, and digital signal processor (DSP) can be applied for the FOC of IM drives [35, 36]. The accuracy and speed of limited processing and long time-delay transmitter are the most important parameters to suitable selection of the computing device for implementing of high performance FOC electrical machines [37].

The development of microcontrollers' technology from 4-bit with limited capabilities in 1971 to complex 64-bit architecture in 1992, enables the implementation of IM drives incorporating flux control scheme [38]. Three main families of microcontroller such as MCS-52, MCS-96, and I960 have developed based on 8-bit and 16-bit complex instruction set computing (CISC), 32bit and 64 bit reduced instruction set computing (RISC) microprocessor architecture, respectively. The developed 32-bit RISC microprocessors have been assisted to implement the fully digital control system for control the nonlinear dynamic IM drives [39, 40]. An electrical motor drive is implemented based on PWM modulation by using PMACP16-200 microcontroller [41]. Another implementation is done by using the Motorola microprocessor MC68HC11E-9 to fully digitally control of IM drive [42].

The DSPs advent came around 1979 and their technology update has continued until now. The advanced DSPs, RISC processors, and parallel processors to the most proposed applications, especially for FOC induction motor drives, provide the high computing capabilities. Considering advances in very large scale integration (VLSI) and very high speed integrated circuit description language (VHDL) technologies, the advanced DSPs are effectively able to realize advancement of the control scheme. The FOC of an IM drive has been implemented by the advanced DSPs [43, 44].

To implement a more complex application, embedded systems (ESs) which are the incorporated computers in computing devices are applied to perform application specific functions. The ESs include a variety of computing devices such as a microcontroller, application specific integrated circuit (ASIC), application specific integrated processors (ASIP) and DSPs. The new DSP TMS320F28335 made by Texas Instruments Company is one of the ESs. This ES is able to implement high performance electric motor drive with the mentioned characteristic [37]. Thus, this DSP is used in this book.

2.5. Speed Controllers for IM Drive

Over the years, researchers have applied linear methods [45, 46], Intelligent methods [47, 48], and various adaptive control techniques for IM drives [49, 50]. Combination of Adaptive and Intelligent methods may also be used depending on the applications [51, 52]. The related literature searches are provided below.

2.5.1. The PI Controller

Generally, the PI controller is the most common approaches for speed control due to some of its advantages such as, (1) it has a relatively simple structure and clear relationship between its parameters and the system response characteristics so that it can be understood and implemented in practice [53, 54]; (2) many advanced algorithms and strategies such as model predictive control can be applied with this controller [55-61]. However, despite its widespread use, this controller alone is not acceptable as a proper speed controller for IM drives [6] as the PI controller is sensitive to the system parameter variations, and unable to reject the external load disturbance [62-64]. The IM drives are nonlinear, time variant, complex, and uncertain systems and system conditions may change during operation, while the PI controller as a linear controller are valid to operate within certain specific range and consequently it is unable to deal with uncertainties. To overcome these problems, recent literature has been paid significant attention to the potential of the fuzzy logic controller (FLC) for the speed controller of electrical motor drives [65, 66].

2.5.2. Fuzzy Logic Controller

The fuzzy set theory which is known as the mathematical tool for FLC was introduced by Zadeh [67]. Over the past three decades, the field of fuzzy controller applications has spread to include many industrial control applications, and significant research work has supported the development of fuzzy controllers. In 1974, Mamdani [68] pioneered the investigation of the feasibility of using the compositional rule of inference that has been proposed by Zadeh [69], for controlling a dynamic plant. A year later, Mamdani and

Assilian [70] developed the first FLC, and it was successfully implemented to control a laboratory steam engine plant. Mamdani's pioneering work also introduced the most common and robust fuzzy reasoning method, called Zadeh–Mamdani min–max gravity reasoning. Also, a significant number of in-depth theoretical and analytical investigations related to this structure has been reported in [71, 72]. Takagi and Sugeno [73] introduced a different linguistic description of the output fuzzy sets, and a numerical optimization approach to design fuzzy controller structures.

As compared to the conventional PI, PID, and their adaptive versions, the FLCs have some advantages such as: (1) they do not need any exact system mathematical model [74]; (2) they can handle nonlinearity of arbitrary complexity[6]; (3) sometimes they prove to be more robust against plant parameter variation, and load disturbance effects than conventional controllers [75]; (4) they have satisfactory performance with imprecision signals from the sensors [76], and (5) they are based on the linguistic rules with an IF–THEN general structure, which is the basis of human logic [77].

In spite of these advantages, FLC has its critics, particularly among the conventional control theorists, commonly mentioned weakness includes the lack of a formal design and analysis methodologies, including the difficulty in obtaining stability and robustness indices [78, 79]. However, the application of FLC has faced some disadvantages during hardware and software implementation due to its high computational burden [66, 80]. To minimize the real-time computational burden of an FLC, a method based on simple membership functions (MFs) and rules has been implemented in [65].

The FLC has emerged as a complement to conventional strict methods. Design objectives, which are mathematically hard to express can be incorporated into FLC by linguistic rules. The FLC have been developed and can be divided into two groups [81]. The first group focuses on improving the design and performance of the standard FLC [65, 81-83]. The second group of approaches combines the advantages of FLC and those of conventional adaptive techniques. In the early years, most FLCs were designed by trial and error. Since the complexity of a FLC will increase exponentially when it is used to control complex systems, it is tedious to design and tune FLCs manually for most industrial problems like control of motors. That is why, the conventional nonlinear design method [84] was adopted in the fuzzy control area, such as fuzzy sliding control [52, 85], fuzzy gain scheduling [86], various forms of self-tuning and self-organizing FLCs [87, 88], and adaptive fuzzy control [74, 89], in order to alleviate difficulties in constructing the fuzzy rule base and improve the performance of the drive under severe perturbations of model parameters and operating conditions. Also to achieve more improved performance and increased robustness, recently artificial neural networks and genetic algorithms are being used in designing such controllers [51, 90-92].

Considering the aforementioned problems of FLC, recently authors paid attention to the design of the fuzzy controller on the basis of the SMC system for motor drives [93-97].

2.5.3. The Sliding-Mode Controller

To control electric drive systems, the SMC system as a robust control technique is an effective way with many attractive features [98-100] as follows: (1) it is robust to parameter variations and model uncertainties and insensitive to external load disturbance [101, 102]; (2) it offers a fast dynamic response, good transient performance, and stable control system [103]; (3) it can handle some nonlinear systems that are not stable by using a linear controller [104]; and (4) it requires an easy hardware/software implementation [105]. However, due to discontinuous control nature, it has some limitations in electrical drives and shows high-frequency oscillations as chattering characteristic. The chattering can excite high frequency unmodeled dynamics of the system and make various undesirable effects such as current harmonics and torque pulsation [106-108].

In recent years, the chattering issue has become the research focus of many scholars [103, 109, 110]. These researches have been done utilizing fuzzy controllers to reduce the chattering due to the advantages of the designed fuzzy controller on the basis of the SMC system [96, 111]. To put more emphasis on this issue, two controllers with the different structure such as, boundary layer fuzzy controller (BLFC) and fuzzy sliding mode controller (FSMC) are pointed out and reviewed in details in the next sections to show the gaps in this area.

2.5.4. Boundary Layer Fuzzy Controller

The fuzzy approach has been used to handle uncertainty in plant dynamics over the past three decades [112-114] as it does not need an accurate mathematical model and it can handle any nonlinearities of arbitrary complexities.

Generally, introducing a thin boundary layer around the sliding surface can solve the chattering problem by interpolating a continuous function inside the boundary layer of switching surface [115-117]. However, in most of the reported works, the slope of the continuous function is a compromise between control performance and chattering elimination. Thus, the chattering in torque response cannot be eliminated with the reported boundary layer approach. In addition, the asymptotic stability is not guaranteed and may cause a steady-state error.

Authors utilized two different fuzzy inference systems to adjust and change the slope of the continuous function to improve the tracking performance considering the thin boundary layer near sliding surface [115, 117]. However, the validity of these works was not tested by experimental results for parameter variations and external load disturbances.

Fuzzy inference systems were applied to design two kinds of nonlinear boundary layers to reduce the chattering [118]. However, the fuzzy inference system was a complicated one with so many fuzzy memberships and rules. Thus, this controller is not suitable for real-time implementation and hence, authors provided only simulations.

To improve the tracking performance considering the thin boundary layer near sliding surface, the slope of the continuous function or boundary layer thickness was adjusted by the fuzzy inference system [109]. However, the authors considered only one input (speed error) which in not sufficient to handle the external large disturbances. Moreover, the controller might get unstable without considering the acceleration as input to the SMC. That is why, authors did not provide any results with large external load disturbances. As IM drive often faces the possibility of large uncertainties including large external load disturbance and variations of critical motor parameters in real-time, the controller may not be suitable for high performance drives. Authors reported that for large disturbances, the controller needs a high gain of reaching control part and a thicker boundary layer to eliminate the chattering effect. However, increase the boundary layer thickness decreases the feedback to a system without sliding mode [105]. Therefore, in this book a novel boundary layer fuzzy controller (NBLFC) is developed for the IM drive to overcome the drawbacks of the existing BLFC controllers [109, 115].

2.5.5. Fuzzy Sliding Mode Controller

Despite many advantages of the fuzzy logic approach, the huge number of fuzzy rules for a high-order system makes the analysis complex. For example, Liaw et al. introduced a two-degree-of-freedom controller with fuzzy adaptation to reduce the effects of parameter variations on the desired performance [119]. However, the adopted linguistic rule base was too complex, which result in a high computational burden and was almost impossible to apply in real-time [97].

Adaptive fuzzy sliding-mode position controllers for an IFOC of IM drive were investigated in [97, 120]. However, the singleton fuzzifiers were used for output fuzzy sets, which not sufficient to cover all operating conditions. Moreover, the authors in these works did not test the performance of the IM drive against step changes of position, speed, and load, which frequently encounter in practice.

An AFSMC control technique was reported by [106]. However, the AFSMC was not free from chattering in torque responses due to the lack in proper development of control algorithm. Thus, the developed AFSMC is not acceptable for high performance drives.

A fuzzy sliding mode speed controller with a load torque observer was designed for a permanent magnet synchronous motor (PMSM) drive [111]. However, the structure of FSMC was not adapted online to cope with uncertainties and minimize the control effort. In addition, the dynamic performance of the proposed controller was almost same as the fixed boundary layer sliding-mode controller (FBLSMC). To overcome the disadvantages of the aforementioned FSMC and the FBLSMC controllers, a new adaptive fuzzy sliding-mode controller (AFSMC) is developed for an IM drive in this book.

2.6. Motivation of the Current Work

From the above literature review, the following conclusions or directions can be drawn.

i. It is pellucid that IM drives along with SVM-voltage–fed converter have been applied widely in industrial applications due to their advantages. Among the scalar and vector control methods, the IFOC based vector control method has been received more attention to control IM drives. Since IFOC with conventional controllers are sensitive to the system parameter variations and insufficient to reject the external load disturbance, the appropriate design of speed controller is necessary to satisfy the high performance and robustness conditions of IM drive.
ii. Variable-structure control (VSC), and in particular, sliding mode control (SMC) system as a robust control technique is an effective

way with many attractive features which is suitable for a high performance IM drive. However, due to discontinuous control nature, it shows high-frequency oscillations as chattering, which can make various undesirable effects. Due to the salient advantages of the fuzzy controller, which has been designed based on the SMC system, the recent focus of researches has been put on the FSMC and BLFC controllers based on the fixed and variable boundary layer SMC, respectively.

iii. In case of a variable boundary layer SMC, authors reported some BLFC controllers to improve the chattering phenomena and dynamic performance of the IM drive in recent years. However, the existing BLFC controllers could not eliminate the chattering in torque and speed responses completely and could not improve the dynamic performance of the IM drive at different operation conditions and uncertainties due to the lack of proper development of those controllers as listed below.
   1) The complexity of the applied fuzzy inference systems is not suitable for real-time implementation.
   2) Compromising between control performances, chattering elimination and also existing the steady-state error due to the presence of a continuous function inside the boundary layer.
   3) Considering only one input (speed error) to the SMC which in not sufficient to handle the external large disturbances and may cause instability.
   4) The possibility of large uncertainties in real-time applications needs a high gain of reaching control part, which reduces the feedback to a system without sliding mode.

Hence, to overcome the disadvantages of existing BLFC, a NBLFC is developed for high performance IM drive in this book.

iv. By incorporation of fuzzy controller with the SMC system, authors reported some FSMC to improve the chattering and dynamic performance of the IM drive in recent years. However, the existing FSMC was unable to eliminate the chattering and improved dynamic performance in torque and speed responses of the IM drive at different operation conditions and uncertainties due to the lack of proper development of those controllers as listed below.

1) Existing the huge number of fuzzy rules and fuzzy membership functions, which result in a high computational burden and was almost impossible to apply in real-time.
2) Using the singleton fuzzifier as output fuzzy sets, which not sufficient to cover all operating conditions such as, step changes of position, speed, and load.
3) The lock of online adaptation in the structure of FSMC to cope with uncertainties and minimize the control effort, which result in the dynamic performance of the FSMC controller was almost same as the BLSMC controller.

Therefore, in the book an attempt is made to develop a novel fuzzy sliding-mode controller (NFSMC) based high performance IM drive, which overcomes the shortcomings of the existing FSMC.

2.7. Theoretical Background on Selection Methods

This section briefly reviews the basic theoretical background on the related theories in the FSMC for a high performance drive including mathematical model of IM for IFOC, the principle of fuzzy logic, and the sliding-mode speed controller, respectively.

2.7.1. Mathematical Model of IM for IFOC

Due to the position changes of the rotor winding with respect to the stator winding, the mutual inductances represented in the induction motor model varies with time, rendering it difficult to apply the same in the drive system model and analysis. By using the vector control, the IM can be represented as two-phase motor in a stationary reference frame (d^s, q^s) and then transform in synchronously rotating reference frame (d^e, q^e) by applying Park's transformation [121].

Figure 2.1 shows the block diagram of the IFOC Induction motor drive.

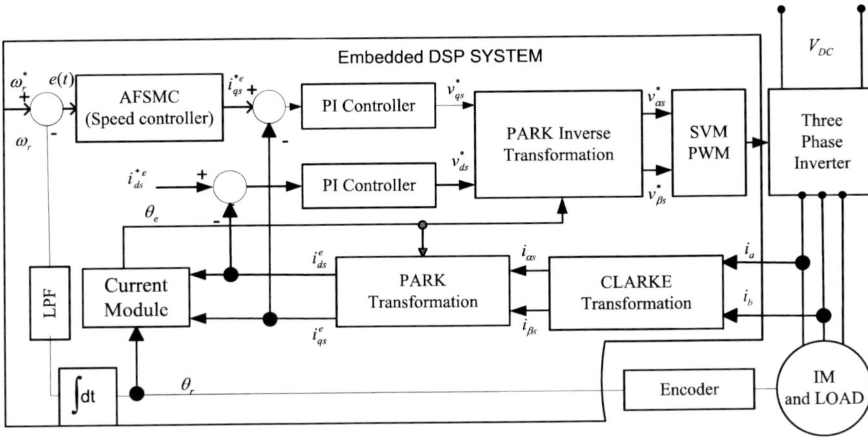

Figure 2.1. Control structure of an IFOC induction motor.

As shown in Figure 2.1, Park's transformation along with its inverse and Clarke's transformation are used to implement the IFOC of IM drive. Equations of Park's transformation and its inverse and Clarke's transformation are shown in (2.1) to 2.3), respectively.

$$x_{dq} = T(x_{\alpha\beta}) = \begin{bmatrix} x_d \\ x_q \end{bmatrix} = \begin{bmatrix} \cos\theta_e & \sin\theta_e \\ -\sin\theta_e & \cos\theta_e \end{bmatrix} \begin{bmatrix} x_\alpha \\ x_\beta \end{bmatrix} \tag{2.1}$$

$$x_{\alpha\beta} = T^{-1}(x_{dq}) = \begin{bmatrix} x_\alpha \\ x_\beta \end{bmatrix} = \begin{bmatrix} \cos\theta_e & -\sin\theta_e \\ \sin\theta_e & \cos\theta_e \end{bmatrix} \begin{bmatrix} x_d \\ x_q \end{bmatrix} \tag{2.2}$$

$$x_{\alpha\beta} = T(x_{abc}) = \begin{bmatrix} x_\alpha \\ x_\beta \end{bmatrix} = \frac{2}{3} \begin{bmatrix} 1 & -1/2 & -1/2 \\ 0 & \sqrt{3}/2 & -\sqrt{3}/2 \end{bmatrix} \begin{bmatrix} x_a \\ x_b \\ x_c \end{bmatrix} \tag{2.3}$$

In the above equations, x can be phase variables of voltage or current so that α phase coordinates with d-axis. Since the Clarke's transformation has two inputs (shown in Figure 2.1) considering balanced three phase induction motor, x_c is replaced with $-(x_a + x_b)$ and the Clarke's transformation is changed as,

$$x_{\alpha\beta} = T(x_{ab}) = \begin{bmatrix} x_\alpha \\ x_\beta \end{bmatrix} = \begin{bmatrix} 1 & 0 \\ 1/\sqrt{3} & 2/\sqrt{3} \end{bmatrix} \begin{bmatrix} x_a \\ x_b \end{bmatrix} \tag{2.4}$$

These concepts assist to model the IM for IFOC as shown in the next section.

2.7.2. Induction Motor Equation

The concept of d-q representations can be utilized to develop the basic machine equations for vector control application in a dynamic model [122, 123]. Equation (2.5) shows, the d-q axes model of an induction motor using reference axis rotating at synchronous speed.

$$\begin{bmatrix} v_{ds}^e \\ v_{qs}^e \\ 0 \\ 0 \end{bmatrix} = \begin{bmatrix} R_s + \sigma L_s p & -\sigma L_s \omega_e & \frac{L_m}{L_r} p & -\frac{L_m}{L_r} \omega_e \\ \sigma L_s \omega_e & R_s + \sigma L_s p & \frac{L_m}{L_r} \omega_e & \frac{L_m}{L_r} p \\ -L_m \frac{R_r}{L_r} & 0 & \frac{R_r}{L_r} + p & -\omega_{sl} \\ 0 & -L_m \frac{R_r}{L_r} & \omega_{sl} & \frac{R_r}{L_r} + p \end{bmatrix} \begin{bmatrix} i_{ds}^e \\ i_{qs}^e \\ \varphi_{dr}^e \\ \varphi_{qr}^e \end{bmatrix} \tag{2.5}$$

where,

$$\sigma = 1 - \frac{L_m^2}{L_s L_r}, p = \frac{d}{dt}, \omega_{sl} = (\omega_e - \omega_r) \tag{2.6}$$

The electromagnetic torque of 3-phase and P-pole induction motor is obtained as,

$$T_e = \frac{3}{2}\frac{P}{2}\frac{L_m^{\ 2}}{L_r}(i_{qs}^e \varphi_{dr}^e - i_{ds}^e \varphi_{qr}^e) \tag{2.7}$$

where,

$$\varphi_{dr}^e = (L_m\, i_{ds}^e + L_r\, i_{dr}^e) \tag{2.8}$$

$$\varphi_{qr}^e = (L_m\, i_{qs}^e + L_r\, i_{qr}^e) \tag{2.9}$$

2.7.3. Indirect Field Oriented Control

The FOC mentions that the stator current components are oriented in flux and torque component in reference axis rotating at synchronous speed. Rotor voltage equations in (2.5) is can be rewritten as,

$$p\varphi_{dr}^e + \frac{R_r}{L_r}\varphi_{dr}^e - \frac{L_m}{L_r}R_r i_{ds}^e - \omega_{sl}\varphi_{qr}^e = 0 \tag{2.10}$$

$$p\varphi_{qr}^e + \frac{R_r}{L_r}\varphi_{qr}^e - \frac{L_m}{L_r}R_r i_{qs}^e + \omega_{sl}\varphi_{dr}^e = 0 \tag{2.11}$$

where,

$$L_{ls} = L_s - L_m, L_{lr} = L_r - L_m \tag{2.12}$$

For vector control, φ_{qr} is set to zero so that the flux is oriented in the d-axis.

$$\varphi_r^e = \varphi_{dr}^e \tag{2.13}$$

$$\varphi_{qr}^e = 0 \tag{2.14}$$

Substituting (2.13) & (2.14) in (2.10) & (2.11) yields,

$$\omega_{sl} = \frac{L_m}{\varphi_r^e}(\frac{R_r}{L_r})i_{qs}^e \tag{2.15}$$

$$\frac{L_r}{R_r} p\varphi_r^e + \varphi_r^e = L_m i_{ds}^e \tag{2.16}$$

When rotor flux is constant under base speed, then

$$p\varphi_{dr}^e = 0 \tag{2.17}$$

Considering the equation (2.9), (2.14) yields,

$$i_{qs}^e = -\frac{L_r}{L_m} i_{qr}^e \tag{2.18}$$

Equation (2.18) shows a direct equilibrium relation between i_{qs}^e , and i_{qr}^e . Equation (2.16) can be shown in the steady state as,

$$\varphi_{dr}^e = L_m i_{ds}^{*e} \tag{2.19}$$

Considering (3.7), (3.14), and (3.19) the electromagnetic torque equation is obtained as,

$$T_e = \frac{3}{2}\frac{P}{2}\frac{L_m{}^2}{L_r} i_{qs}^{*e} i_{ds}^{*e} \tag{2.20}$$

The electromagnetic torque can be simplified as,

$$T_e = K_t i_{qs}^{*e} \tag{2.21}$$

where,

$$K_t = \frac{3}{2}\frac{P}{2}\frac{L_m{}^2}{L_r} i_{ds}^{e*} \tag{2.22}$$

According to (2.21), with the flux command held constant, a change in i_{qs}^{*e} instantly results in a corresponding change in i_{qr}^{e}. From (2.19) and (2.21), it is easily seen that the flux and torque vector of IM are decoupled. According to (2.15), an exact estimation of slip frequency requires the proper knowledge of rotor time constant $T_r = L_r / R_r$ that concludes suitable estimation of position stator.

Considering the implementation of IFOC, the induction motor drive can be simplified as shown in Figure 2.2 and the mechanical equation of an induction motor can be presented as follows [123].

$$J_r \dot{\omega}_r(t) + B\omega_r(t) = T_e - T_L \tag{2.23}$$

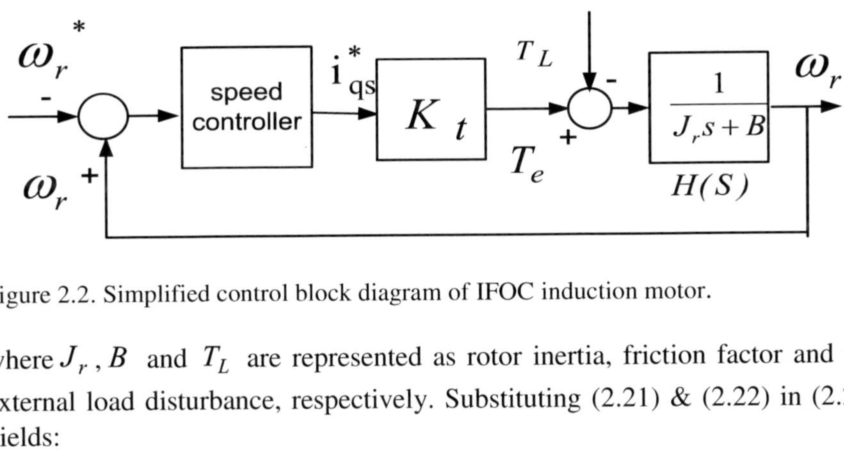

Figure 2.2. Simplified control block diagram of IFOC induction motor.

where J_r, B and T_L are represented as rotor inertia, friction factor and the external load disturbance, respectively. Substituting (2.21) & (2.22) in (2.23) yields:

$$\begin{aligned}\dot{\omega}_r(t) &= -\frac{B}{J_r}\omega_r(t) + \frac{k_t}{J_r} i_{qs}^{*e} - \frac{T_L}{J_r} \\ &= B_p\omega_r + A_p i_{qs}^{*e} + D_p T_L\end{aligned} \tag{2.24}$$

where, $A_p = k_t / J_r > 0$, $B_p = -B / J_r < 0$ and $D_p = -1 / J_r > 0$.

To achieve the nominal model of IM drive, nominal value of the parameters must be considered without any disturbances. Thus, the nominal model of the IM drive given by (3.24) can be written as,

$$\dot{\omega}_r(t) = \overline{B}\omega_r + \overline{A}i_{qs}^{*e} \tag{2.25}$$

where, $\overline{A} = \overline{k_t} / \overline{J_r}$ and $\overline{B} = -\overline{B} / \overline{J}_r$ are the nominal values of A_p and B_p, respectively. To handle the uncertainties, they must be considered and added to the nominal model for real-time induction motor (IM) drive [124]. So, the dynamic equation (2.25) considering structured and unstructured uncertainties and the unmodeled dynamics for the actual induction motor drive can be written as follows:

$$\begin{aligned}\dot{\omega}_r(t) &= (\overline{B} + \Delta B)\omega_r(t) + (\overline{A} + \Delta A)i_{qs}^{*e} + D_p T_L + \delta \\ &= \overline{B}\omega_r(t) + \overline{A}i_{qs}^{*e} + L(t)\end{aligned} \tag{2.26}$$

In (3.26), the uncertainties are shown by ΔA & ΔB. In addition, unstructured uncertainty due to detuning field-orientation in the transient state and the unmodeled dynamics in practical applications are shown as δ. In the equation above, $L(t)$ is called lumped uncertainty and it is obtained as

$$L(t) = \Delta B\omega_r(t) + \Delta A i_{qs}^{*e} + D_p T_L + \delta \tag{2.27}$$

2.8. The Principle of Fuzzy Logic

The fuzzy approach as a control methodology is able to processing data to deal with uncertain systems by using linguistic terms and degree of memberships. This method was not applied until 1970 due to insufficient computing capability of the processor. Due to the rapid increase in computer technology, the applications and development of the fuzzy systems have been extensively used to control the plants in the industry [87, 88, 125-129]. Fuzzy controllers act like human logic which use the linguistic terms to express uncertain quantities [130]. Investigation of Mamdani in 1974 leads to a Mamdani fuzzy model. The fuzzy inference system (FIS) includes four parts such as fuzzification, interface engine, fuzzy rule base, and defuzzification [68] as shown in Figure 2.3.

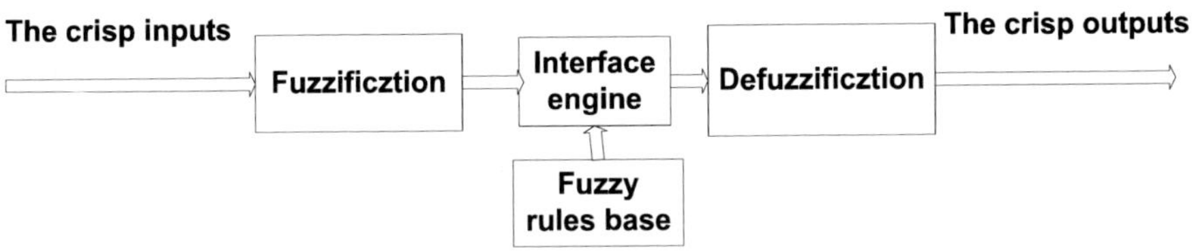

Figure 2.3. The Mamdani's fuzzy system inference.

In the FIS system, each fuzzy number is declared by a fuzzy set and requires a specified range of crisp with grade of membership. The fuzzy sets of inputs MFs transfer crisp inputs into fuzzy inputs (fuzzification). To express fuzzy sets from a mathematical point of view, consider a set of objects X. The set is explained as follows:

$$X = x_1, x_2, ..., x_n \tag{2.28}$$

where, x_i is an element in the set X. A membership value (μ) expresses the grade of membership related to each element x_i in a fuzzy set A, which shows a combination as below:

$$A = \mu_1(x_1), \mu_2(x_{2)}, ..., \mu_n(x_n) \tag{2.29}$$

The membership function as inputs and outputs of a fuzzy system presents participation amplitude of each input and output with several types such as linear (trapezoidal, triangular), concave, and exponential type. It is noted that the MFs are applied in the trapezoidal and triangular forms in this book. A trapezoidal fuzzy number can be shown as,

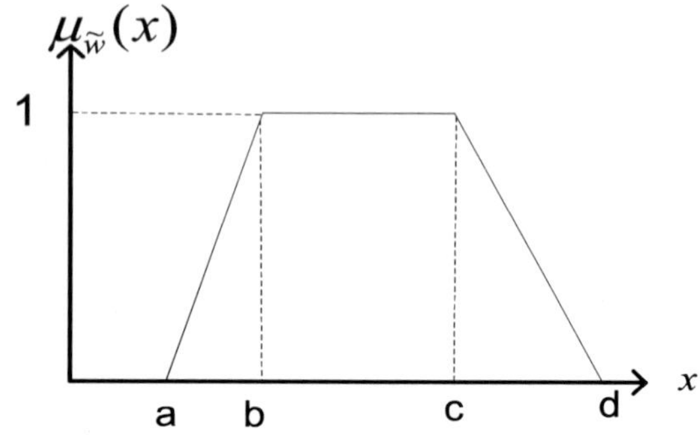

Figure 2.4. The trapezoidal fuzzy membership function.

$$\tilde{w} = (a, b, c, d) \tag{2.30}$$

The trapezoidal MF is defined in (3.30) and as shown in Figure 3.4. According to (2.27), if $b = c$ then the number is called a triangular fuzzy number.

$$\mu_{\tilde{w}}(x) = \begin{cases} 0 & if \quad x < a \\ \frac{1}{b-a}(x-a) & if \quad a \le x \le b \\ 1 & if \quad b \le x \le c \\ \frac{1}{c-d}(x-d) & if \quad c \le x \le d \\ 0 & if \quad x > d \end{cases} \tag{2.31}$$

Rules of the FIS system as a main part of the system based on experts' knowledge in each area are available in the form of "IF-THEN." For instance, "IF x_1 is a_1 and x_2 is b_1, THEN y is c_1" so that x_1 and x_2 are variables, y is a solution variable, and a_1, b_1, and c_1 are fuzzy linguistic terms. The interface engine of the FIS system does the integration of fuzzy sets based on the fuzzy rules and allocation of these integrations in the special related fuzzy area. Finally, in defuzzification process, fuzzy number transforms to crisp output. The center of area method (COA), the bisector of area method (BOA), mean of maximum method (MOM), smallest of maximum method (SOM), and the largest of maximum method (LOM) are most used common method of deffuzification [131]. In this book, the COA method is used for defuzzification as shown in (2.32).

$$x_{COA} = \frac{\sum_{i=1}^{n} x_i . \mu_i(x_i)}{\sum_{i=1}^{n} \mu_i(x_i)} \tag{2.32}$$

where, x_i is an element in the set X as mentioned in (3.28) and (3.29).

2.9. The Sliding-Mode Speed Controller

To force the motor to follow the command speed, a speed controller is applied as shown in Figure 3.1. Generally, the speed error between the commanded speed and actual speed is used as input to this controller and the output of the speed controller is q-axis command stator current. The reference stator current (i_{qs}^{*e}) provides the required two axis command voltage (v_{qs}^{*e}, v_{ds}^{*e}) to keep the desired speed through two PI controllers. The SVM based voltage source inverter produces the required voltage to drive the motor. The accuracy of speed controller plays an important role for high performance drives. Therefore, in this book the sliding-mode and fuzzy logic approaches are considered to develop the speed controller.

2.9.1. Sliding –Mode Controller Theory

2.9.1.1. Introduction

In case of control, an uncertain nonlinear system depending on its uncertainties can be classified into two types. In the first type, some parameters of the model are unknown or have a variation, which is named structured or parametric uncertainties. In the second one, some parameters cannot be modeled which is called unstructured uncertainties or unmodeled dynamics [132]. Two types of uncertain systems in practical design need to use a robust adaptive control, which is able to handle the uncertainties. The IFOC induction motor drive has both types of uncertainties due to parameter variations, i.e., inaccurate parameters, and unmodeled dynamics. The robust adaptive control is an adaptive control, which includes a nominal part to deal with nominal model and additional terms so that model uncertainties are handled by the additional terms.

Sliding-mode control as an adaptive robust control shows suitable performance in the face of modeling imprecisions while guarantees stability of the system. The SMC system has been applied to several applications such as, robot manipulators, underwater vehicles, automotive transmissions and engines, high performance electrical drive, and power system. In the next sub-section, the phrases of sliding mode are discussed.

2.9.1.2. Sliding Surface and Equivalent Control

A dynamic equation of an n-order uncertain system can be written as,

$$x^{(n)} = f(x) + b(x) * u(t) \tag{2.33}$$

In the above equation, x is scalar variable which is the objective of control, scalar u is controlling input, $f(x)$ and $b(x)$ are imprecise functions owing to the dynamic model, and the state variable is obtained as,

$$X = [x, \dot{x}, ..., x^{n-1}]^T \tag{2.34}$$

The control target can be defined achieving state variable X to the desired variable X_d (2.35) despite uncertainties in the system so that initial condition equation (2.36) is satisfied.

$$X_d = [x_d, \dot{x}_d, ..., x_d^{n-1}]^T \tag{2.35}$$

$$X(0) = X_d(0) \tag{2.36}$$

When initial condition is not satisfied, proper tracking of state variable is done only for steady state.

The suitable choicc of a scalar variable assists in proper tracking of desired state according to the model and its parameter that is to be controlled. Considering the equation (2.37), scalar variable $S(t)$ can be selected [132] in (2.38) which is named the sliding surface.

$$\widetilde{X} = X - X_d = [\widetilde{x}, \dot{\widetilde{x}}, ..., \widetilde{x}^{n-1}]^T \tag{2.37}$$

$$S(X,t) = (\frac{d}{dt} + C)^{n-1} \widetilde{X} \tag{2.38}$$

where, $\widetilde{X}$ is tracking error and C is positive constant. From equation (2.38), it can be seen that in spite of suitable tracking, error $\widetilde{X}$ will be equal to zero and consequently scalar value of S will be equal to zero. In fact, tracking problem involving the n-dimensional vector $\widetilde{X}$ can be reduced to a scalar variable S. Moreover, bounds of variable switching directly transform into the bounds of tracking error. A similar condition can be achieved by using integral switching surface so that the integral of switching variable is replaced by switching variable in the equation (2.38). Therefore, the equation of sliding surface integral with regard to this variable will be equal to the equation (2.39).

$$S(X,t) = (\frac{d}{dt} + C)^{n-1} \int_0^r \widetilde{X} dr \tag{2.39}$$

In the SMC system, dynamic movement of the system on both sides of the sliding surface can be averaged so that an expression is obtained from the control input. So, by solving the dynamic equation of sliding surface as shown in (3.40) an equation can be derived which is called equivalent control.

$$\dot{S}(t) = 0 \tag{2.40}$$

2.9.1.3. Stability Condition and Reaching Phase

By defining a Lyapunov function as,

$$V = \frac{1}{2} S^2 \tag{2.41}$$

Stability condition can be obtained from Lyapunov stability theorem as [132].

$$\dot{V} = S(t)\,\dot{S}(t) \le -\eta |S(t)| \tag{2.42}$$

where, η is a strictly positive constant.

The stability condition (3.42) guarantees stability of a system considering uncertainties in system by keeping the scalar $S(t)$ at zero and selection $u(t)$ as control input from (2.33). In other words, the stability condition makes surface $S(t)$ as an invariant set such that disturbances of dynamic system tolerated around this surface. In this situation, the system that acts with the previously mentioned condition is named sliding-mode system. When a sliding - mode system satisfies the stability condition (2.42) while initial condition equation (2.36) is not exactly verified or defined, tracking is done after finite time as shown in equation (2.43) [132] which is called reaching phase.

$$t_{reach} \le |S(t=0)| / \eta \tag{2.43}$$

In case of induction motor drive considering speed as control target, the tracking error is defined as the difference between actual speed of motor and command speed. In addition, drive dynamic equation is considered second-order system. Moreover, control input $u(t)$ can be selected as reference electrical motor torque T_e^* or the q-axis stator current i_{qs}^{*e}.

2.9.1.4. The Conventional SMC system

Considering the equation (2.44) as speed tracking error and equation (3.38), time-varying surface of sliding mode in the state of space $\Re^2$ is introduced as shown in equation (2.45).

$$e(t) = \omega_r(t) - \omega_r^*(t) \tag{2.44}$$

$$S(t) = h(Ce(t) + \dot{e}(t)) \tag{2.45}$$

where, h in the above scalar equation is a positive constant and $h < 1$. Substituting (2.26), (2.44), and (2.45) in (2.40) yields,

$$\dot{S}(t) = h(C\dot{e}(t) + \overline{B}\dot{\omega}_r(t) + \overline{A}u(t) + \dot{L}(t) - \ddot{\omega}_r^*(t)) \tag{2.46}$$

where, $u(t) = \dot{i}_{qs}^{*e}\, t)$.

In the equation above, $u(t)$ is called the control effort or control law. In (2.46) it is assumed to be the bound of $\dot{L}(t)$ which is unknown but it is limited as,

$$\left|\dot{L}(t)\right| < M \tag{2.47}$$

where, M is positive constant. By solving $\dot{S}(t)$ without consideration of lumped uncertainty ($\dot{L}(t) = 0$), the desired performance under nominal system model can be achieved. In this situation $u(t)$ equals $u_{eq}(t)$ and can be obtained using equation (2.46).

$$u_{eq}(t) = -\overline{A}^{-1}[(C + \overline{B})\dot{e}(t) + \overline{B}\dot{\omega}_r^* - \ddot{\omega}_r^*] \tag{2.48}$$

where, $u_{eq}(t)$ is called the equivalent control part of the control effort. In order to have suitable performance despite uncertainties on the dynamic of system (lumped uncertainty), discontinuous term must be added to the equivalent control across the sliding surface $S(t)$. The term discontinuous is called hitting control part or reaching control part of the control effort [133]. It is given as,

$$u_r(t) = -(\bar{A}h)^{-1} k(t)\,\mathrm{sgn}(S(t)) \tag{2.49}$$

where, $k(t)$ is called the control gain, and Sgn is the sign function as follows.

$$\mathrm{sgn}(S(t)) = \begin{cases} 1 & \textit{if} \quad S(t) > 0 \\ -1 & \textit{if} \quad S(t) < 0 \end{cases}$$

Therefore, the control law is obtained as shown in equation (2.49) and (2.50).

$$u(t) = u_{eq}(t) + u_r(t) \tag{2.50}$$

$$i_{qs}^* = \frac{1}{\tau}\int_0^t u(t)dt \tag{2.51}$$

where, τ is the integral constant.

The stability condition (2.42) guarantees stability of the system considering lumped uncertainties by keeping the scalar $S(t)$ at zero. Then, substituting (2.46) in (2.42) and considering (2.45), stability condition is obtained as,

$$\begin{aligned} \dot{V} &= S(t)\,\dot{S}(t) = -S(t)h\bar{A}u_r(t) + hS(t)\,\dot{L}(t) \\ &\le -k|S(t)| + h|S(t)|\left|\dot{L}(t)\right| \\ &\le -|S(t)|(k(t) - hM) \end{aligned} \tag{2.52}$$

A comparison between (3.41) & (3.51) yields,

$$k(t) \ge hM + \eta \tag{2.53}$$

Then by choosing (2.53), stability of the system (2.50) is guaranteed.

REFERENCES

[1] N. Tesla and D. H. Childress, *The fantastic inventions of Nikola Tesla*: Adventures Unlimited Press, 1993.

[2] D. B. Hoseason, "Squirrel-cage induction motors," *Electrical Engineers, Journal of the Institution of,* vol. 66, pp. 410-425, 1928.

[3] U. A. Bakshi and V. U. Bakshi, *Electrical technology*: Technical Publications, 2009.

[4] F. Blaschke, "The principle of field orientation as applied to the new transvektor closed-loop control system for rotating-field machines," *Adjustable Speed AC Drive Systems,* 1972.

[5] M. N. Uddin and H. Wen, "Development of a self-tuned neuro-fuzzy controller for induction motor drives," *Conference Record of the 2004 IEEE Industry Applications Conference, Vols 1-4,* pp. 2630-2636, 2004.

[6] A. Saghafinia, H. Wooi Ping, and M. Rahman, "High Performance Induction Motor Drive using Hybrid Fuzzy-PI and PI Controllers: a Review," *International Review of Electrical Engineering-Iree,* vol. 5, pp. 2000-2012, october-september 2010.

[7] X. D. Sun, K. H. Koh, B. G. Yu, and M. Matsui, "Fuzzy-Logic-Based $ V/f $ Control of an Induction Motor for a DC Grid Power-Leveling System Using Flywheel Energy Storage Equipment," *Industrial Electronics, IEEE Transactions on,* vol. 56, pp. 3161-3168, 2009.

[8] K. S. Gaeid, H. W. Ping, M. K. Masood, and M. A. Saghafinia, "Induction motor fault tolerant control with wavelet indicator," in *Transportation, Mechanical, and Electrical Engineering (TMEE), 2011 International Conference on*, 2011, pp. 949-953.

[9] W. L. Erdman and R. G. Hoft, "Induction machine field orientation along airgap and stator flux," *IEEE Transaction on Energy Conversion,* vol. 5, pp. 115-121, 1990.

[10] M. Boussak and K. Jarray, "A high-performance sensorless indirect stator flux orientation control of induction motor drive," *Industrial Electronics, IEEE Transactions on,* vol. 53, pp. 41-49, 2006.

[11] S. Vaez-Zadeh and E. Jalali, "Combined vector control and direct torque control method for high performance induction motor drives," *Energy conversion and management,* vol. 48, pp. 3095-3101, 2007.

[12] M. N. Uddin and H. Wen, "Development of a self-tuned neuro-fuzzy controller for induction motor drives," 2004.

[13] B. K. Bose, "Scalar decoupled control of induction motor," *Industry Applications, IEEE Transactions on,* pp. 216-225, 1984.

[14] R. W. De Doncker and D. W. Novotny, "The universal field oriented controller'. Confermce Record IEEE Industry Applications sociay," 1988, pp. 450-456.

[15] S. Ogasawara, H. Akagi, and A. Nabae, "The generalized theory of indirect vector control for AC machines," *Ieee Transactions on Industry Applications,* vol. 24, pp. 470-478, 1988.

[16] D. S. Zinger, F. Profumo, T. Lipo, and D. W. Novotny, "A direct field-oriented controller for induction motor drives using tapped stator windings," *Ieee Transactions on Power Electronics,* vol. 5, pp. 446-453, 1990.

[17] H. Rehman and R. Dhaouadi, "A fuzzy learning--Sliding mode controller for direct field-oriented induction machines," *Neurocomputing,* vol. 71, pp. 2693-2701, 2008.

[18] K. Matsuse, T. Yoshizumi, S. Katsuta, and S. Taniguchi, "High-response flux control of direct-field-oriented induction motor with high efficiency taking core loss into account," *Industry Applications, IEEE Transactions on,* vol. 35, pp. 62-69, 1999.

[19] R. J. Wai and K. M. Lin, "Robust decoupled control of direct field-oriented induction motor drive," *Industrial Electronics, IEEE Transactions on,* vol. 52, pp. 837-854, 2005.

[20] C. Chakraborty and Y. Hori, "Fast efficiency optimization techniques for the indirect vector-controlled induction motor drives," *Industry Applications, IEEE Transactions on,* vol. 39, pp. 1070-1076, 2003.

[21] J. W. Finch and D. Giaouris, "Controlled AC electrical drives," *Industrial Electronics, IEEE Transactions on,* vol. 55, pp. 481-491, 2008.

[22] J. Maes and J. A. Melkebeek, "Speed-sensorless direct torque control of induction motors using an adaptive flux observer," *Industry Applications, IEEE Transactions on,* vol. 36, pp. 778-785, 2000.

[23] J. Holtz and J. Quan, "Sensorless vector control of induction motors at very low speed using a nonlinear inverter model and parameter identification," *Industry Applications, IEEE Transactions on,* vol. 38, pp. 1087-1095, 2002.

[24] S. Bolognani, L. Peretti, and M. Zigliotto, "Parameter Sensitivity Analysis of an ImprovedOpen-Loop Speed Estimate forInduction Motor Drives," *Power Electronics, IEEE Transactions on,* vol. 23, pp. 2127-2135, 2008.

[25] L. Zhen and L. Xu, "On-line fuzzy tuning of indirect field-oriented induction machinedrives," *Ieee Transactions on Power Electronics,* vol. 13, pp. 134-141, 1998.

[26] M. Masiala, B. Vafakhah, J. Salmon, and A. M. Knight, "Fuzzy Self-Tuning Speed Control of an Indirect Field-Oriented Control Induction Motor Drive," in *41st Annual Meeting of the IEEE-Industry-Applications-Society*, Tampa, FL, 2006, pp. 1732-1740.

[27] K. Corzine, S. Sudhoff, and C. Whitcomb, "Performance characteristics of a cascaded two-level converter," *Energy Conversion, IEEE Transactions on,* vol. 14, pp. 433-439, 1999.

[28] A. Schonung and H. Stemmler, "Static frequency changers with subharmonic control in conjunction with reversible variable speed ac drives," *Brown Boveri Rev,* vol. 51, pp. 555–577, 1964.

[29] G. Pfaff, A. Weschta, and A. F. Wick, "Design and experimental results of a brushless ac servo drive," *Industry Applications, IEEE Transactions on,* pp. 814-821, 1984.

[30] H. W. van der Broeck, H. C. Skudelny, and G. V. Stanke, "Analysis and realization of a pulsewidth modulator based on voltage space vectors," *Industry Applications, IEEE Transactions on,* vol. 24, pp. 142-150, 1988.

[31] M. D. Manjrekar, P. K. Steimer, and T. A. Lipo, "Hybrid multilevel power conversion system: A competitive solution for high-power applications," *Industry Applications, IEEE Transactions on,* vol. 36, pp. 834-841, 2000.

[32] B. K. Bose, *Modern power electronics and AC drives*: Prentice Hall PTR USA, 2002.

[33] Z. Zhang, R. Tang, B. Bai, and D. Xie, "Novel Direct Torque Control Based on Space Vector Modulation With Adaptive Stator Flux Observer for Induction Motors," *Magnetics, IEEE Transactions on,* vol. 46, pp. 3133-3136, 2010.

[34] V. Kinnares and C. Charumit, "Modulating functions of space vector PWM for three-leg VSI-fed unbalanced two-phase induction motors," *Power Electronics, IEEE Transactions on,* vol. 24, pp. 1135-1139, 2009.

[35] R. Gabriel, W. Leonhard, and C. J. Nordby, "Field-Oriented Control of a Standard AC Motor Using Microprocessors," *Industry Applications, IEEE Transactions on,* vol. IA-16, pp. 186-192, 1980.

[36] R. Kumar, R. Gupta, and S. Bhangale, "Microprocessor/Digital Control And Artificial Intelligent Vector Control Techniques For Induction Motor Drive," *IETECH Journal of Electrical Analysis,* 2008.

[37] Y. Yanjie, X. Yunlong, C. Jian, and D. Zhiqiang, "The design of excitation control system based on DSP," in *Electronic Measurement & Instruments, 2009. ICEMI '09. 9th International Conference on*, 2009, pp. 2-637-2-641.

[38] P. C. Sen, J. C. Trezise, and M. Sack, "Microprocessor Control of an Induction Motor with Flux Regulation," *Industrial Electronics and Control Instrumentation, IEEE Transactions on,* vol. IECI-28, pp. 17-21, 1981.

[39] K. Yau-Tze and L. Change-Huan, "Analysis and design of microprocessor-based vector-controlled induction motor drives," *Industrial Electronics, IEEE Transactions on,* vol. 39, pp. 46-54, 1992.

[40] A. Brickwedde, "Microprocessor-Based Adaptive Speed and Position Control for Electrical Drives," *Industry Applications, IEEE Transactions on,* vol. IA-21, pp. 1154-1161, 1985.

[41] J. da Costa, H. Câmara, and E. Carati, "A microprocessor based prototype for electrical machines control using PWM modulation," 2003, pp. 1083-1088 vol. 2.

[42] S. M. Bashi, I. Aris, and S. Hamad, "Development of single phase induction motor adjustable speed control using M68HC11E-9 microcontroller," *Journal of Applied Sciences,* vol. 5, pp. 249-252, 2005.

[43] M. N. Uddin and S. W. Nam, "New online loss-minimization-based control of an induction motor drive," *Power Electronics, IEEE Transactions on,* vol. 23, pp. 926-933, 2008.

[44] B. Garcia and A. Ricardo, "Simulation and implementation of vector control of induction motors," 2011.

[45] A. Miloudi and A. Draou, "Variable gain PI controller design for speed control and rotor resistance estimation of an indirect vector controlled induction machine drive," 2002, pp. 323-328 vol. 1.

[46] R. Sepulchre, T. Devos, F. Jadot, and F. Malrait, "Antiwindup design for induction motor control in the field weakening domain," 2011.

[47] R. Arulmozhiyal and K. Baskaran, "Implementation of a Fuzzy PI Controller for Speed Control of Induction Motors Using FPGA," *Journal of Power Electronics,* vol. 10, pp. 65-71, Jan.

[48] M. A. Fnaiech, F. Betin, G. A. Capolino, and F. Fnaiech, "Fuzzy logic and sliding-mode controls applied to six-phase induction machine with

open phases," *Industrial Electronics, IEEE Transactions on,* vol. 57, pp. 354-364, 2010.

[49] S. Maiti, C. Chakraborty, Y. Hori, and M. C. Ta, "Model reference adaptive controller-based rotor resistance and speed estimation techniques for vector controlled induction motor drive utilizing reactive power," *Industrial Electronics, IEEE Transactions on,* vol. 55, pp. 594-601, 2008.

[50] M. Zerikat and S. Chekroun, "Adaptation Learning Speed Control for a High-Performance Induction Motor Using Neural Networks," *Int. J. Signal Syst. Control Eng. Appi,* vol. 2, pp. 15-21, 2009.

[51] T. Orlowska-Kowalska and M. Dybkowski, "Performance analysis of the sensorless adaptive sliding-mode neuro-fuzzy control of the induction motor drive with MRAS-type speed estimator," *Bulletin of the Polish Academy of Sciences: Technical Sciences,* vol. 60, pp. 61-70, 2012.

[52] S. M. Gadoue, D. Giaouris, and J. W. Finch, "MRAS Sensorless Vector Control of an Induction Motor Using New Sliding-Mode and Fuzzy-Logic Adaptation Mechanisms," *Energy Conversion, IEEE Transactions on,* vol. 25, pp. 394-402, 2010.

[53] R. Sepulchre, T. Devos, F. Jadot, and F. Malrait, "Antiwindup Design for Induction Motor Control in the Field Weakening Domain," *Control Systems Technology, IEEE Transactions on,* vol. PP, pp. 1-15, 2011.

[54] G. Kenne, T. Ahmed-Ali, F. Lamnabhi-Lagarrigue, and A. Arzande, "Real-Time Speed and Flux Adaptive Control of Induction Motors Using Unknown Time-Varying Rotor Resistance and Load Torque," *Energy Conversion, IEEE Transactions on,* vol. 24, pp. 375-387, 2009.

[55] P. Viljamaa, "Fuzzy Gain Scheduling and Tuning of Multivariable Fuzzy Control—Methods of Fuzzy Computing in Control Systems," ed: Citeseer, 2002.

[56] S. Tzafestas and N. P. Papanikolopoulos, "Incremental fuzzy expert PID control," *Ieee Transactions on Industrial Electronics,* vol. 37, pp. 365-371, 1990.

[57] W. Gou-Jen, F. Chuan-Tzueng, and K. J. Chang, "Neural-network-based self-tuning PI controller for precise motion control of PMAC motors," *Industrial Electronics, IEEE Transactions on,* vol. 48, pp. 408-415, 2001.

[58] B. Singh and S. G. Choudhuri, "Fuzzy logic based speed controllers for vector controlled induction motor drive," *Iete Journal of Research,* vol. 48, pp. 441-447, Nov-Dec 2002.

[59] S. V. Ustun and M. Demirtas, "Optimal tuning of PI coefficients by using fuzzy-genetic for V/f controlled induction motor," *Expert Systems with Applications,* vol. 34, pp. 2714-2720, 2008.

[60] S. Z. He, S. Tan, F. L. Xu, and P. Z. Wang, "Fuzzy self-tuning of PID controllers," *Fuzzy Sets and Systems,* vol. 56, pp. 37-46, 1993.

[61] S. Jafarzadeh, C. Lascu, and S. Fadali, "State Estimation of Induction Motor Drives Using the Unscented Kalman Filter," *Industrial Electronics, IEEE Transactions on,* vol. PP, pp. 1-1, 2011.

[62] M. S. Zaky, "Stability Analysis of Speed and Stator Resistance Estimators for Sensorless Induction Motor Drives," *Industrial Electronics, IEEE Transactions on,* vol. 59, pp. 858-870, 2012.

[63] S. Hwi-Beom and P. Jong-Gyu, "Anti-Windup PID Controller With Integral State Predictor for Variable-Speed Motor Drives," *Industrial Electronics, IEEE Transactions on,* vol. 59, pp. 1509-1516, 2012.

[64] F. R. Salmasi and T. A. Najafabadi, "An Adaptive Observer With Online Rotor and Stator Resistance Estimation for Induction Motors With One Phase Current Sensor," *Energy Conversion, IEEE Transactions on,* vol. 26, pp. 959-966, 2011.

[65] M. N. Uddin, T. S. Radwan, and M. A. Rahman, "Performances of fuzzy-logic based indirect vector control for induction motor drive," *IEEE Transactions on Industry Applications,* vol. 38, pp. 1219-1225, Sep-Oct 2002.

[66] M. Singh and A. Chandra, "Application of adaptive network-based fuzzy inference system for sensorless control of PMSG-based wind turbine with nonlinear-load-compensation capabilities," *Power Electronics, IEEE Transactions on,* vol. 26, pp. 165-175, 2011.

[67] L. A. Zadeh, "Fuzzy sets*," *Information and control,* vol. 8, pp. 338-353, 1965.

[68] E. H. Mamdani, "Application of fuzzy algorithms for control of simple dynamic plant," *Proc. Iee,* vol. 121, pp. 1585-1588, 1974.

[69] L. A. Zadeh, "Outline of a New Approach to the Analysis of Complex Systems and Decision Processes," *Systems, Man and Cybernetics, IEEE Transactions on,* vol. SMC-3, pp. 28-44, 1973.

[70] E. H. Mamdani and S. Assilian, "An experiment in linguistic synthesis with a fuzzy logic controller," *International Journal of Man-Machine Studies,* vol. 7, pp. 1-13, 1975.

[71] M. Braae and D. A. Rutherford, "Selection of parameters for a fuzzy logic controller," *Fuzzy Sets and Systems,* vol. 2, pp. 185-199, 1979.

[72] M. Mizumoto, "Realization of PID controls by fuzzy control methods," *Fuzzy Sets and Systems,* vol. 70, pp. 171-182, 1995.

[73] T. Takagi and M. Sugeno, "Fuzzy identification of systems and its applications to modeling and control," *IEEE transactions on systems, man, and cybernetics,* vol. 15, pp. 116-132, 1985.

[74] G. G. Rigatos, "Adaptive fuzzy control for field-oriented induction motor drives," *Neural Computing & Applications,* pp. 1-15, 2012.

[75] R. Palm, "Sliding mode fuzzy control," 1992, pp. 519-526.

[76] J. Zhao, B. K. Bose, and Ieee, "Membership function distribution effect on fuzzy logic controlled induction motor drive," in *Iecon'03: The 29th Annual Conference of the Ieee Industrial Electronics Society, Vols 1 - 3, Proceedings*, ed New York: Ieee, 2003, pp. 214-219.

[77] M. Sugeno, *Industrial applications of fuzzy control*: Elsevier Science Inc. New York, NY, USA, 1985.

[78] G. C. D. Sousa and B. K. Bose, "Fuzzy logic applications to power electronics and drives-an overview," in *Industrial Electronics, Control, and Instrumentation, 1995., Proceedings of the 1995 IEEE IECON 21st International Conference on*, 1995, pp. 57-62 vol.1.

[79] N. Islam, M. Haider, M. B. Uddin, and Ieee, "Fuzzy logic enhanced speed control system of a VSI-fed three phase induction motor," in *2nd International Conference on Electrical and Electronics Engineering (ICEEE 2005)*, Mexico City, MEXICO, 2005, pp. 296-301.

[80] S. Bolognani and M. Zigliotto, "Hardware and software effective configurations for multi-input fuzzy logic controllers," *Fuzzy Systems, IEEE Transactions on,* vol. 6, pp. 173-179, 1998.

[81] M. Masiala, B. Vafakhah, J. Salmon, and A. M. Knight, "Fuzzy Self-Tuning Speed Control of an Indirect Field-Oriented Control Induction Motor Drive," *Industry Applications, IEEE Transactions on,* vol. 44, pp. 1732-1740, 2008.

[82] F. Cupertino, A. Lattanzi, and L. Salvatore, "A new fuzzy logic-based controller design method for DC and AC impressed-voltage drives," *Power Electronics, IEEE Transactions on,* vol. 15, pp. 974-982, 2000.

[83] L. Zhen and L. Y. Xu, "On-line fuzzy tuning of indirect field-oriented induction machine drives," *Ieee Transactions on Power Electronics,* vol. 13, pp. 134-141, 1998.

[84] R. J. Wai and K. H. Su, "Adaptive enhanced fuzzy sliding-mode control for electrical servo drive," *Ieee Transactions on Industrial Electronics,* vol. 53, pp. 569-580, 2006.

[85] L. Li and Z. Xizheng, "Regular paper Indirect Adaptive Fuzzy Sliding-mode Control for Induction Motor Drive," *J. Electrical Systems,* vol. 7, pp. 412-422, 2011.

[86] E. Cerruto, A. Consoli, A. Raciti, and A. Testa, "Fuzzy adaptive vector control of induction motor drives," *Power Electronics, IEEE Transactions on,* vol. 12, pp. 1028-1040, 1997.

[87] A. Saghafinia, H. Ping, and M. Uddin, "Designing Self-Tuning Mechanism On Hybrid Fuzzy Controller For High Performance And Robust Induction Motor Drive," *the International Journal of Advanced Technology & Engineering Research,* vol. 3, 2013.

[88] A. Saghafinia and H. W. Ping, "High performance induction motor drive using fuzzy self-tuning hybrid fuzzy controller," in *Power and Energy (PECon), 2010 IEEE International Conference on*, 2010, pp. 468-473.

[89] S. Y. Wang, C. L. Tseng, and C. J. Chiu, "Design of adaptive TSK-fuzzy observer for vector control induction motor drives," 2011, pp. 5220-5223.

[90] M. N. Uddin and W. Hao, "Development of a Self-Tuned Neuro-Fuzzy Controller for Induction Motor Drives," *Industry Applications, IEEE Transactions on,* vol. 43, pp. 1108-1116, 2007.

[91] M. Cirrincione, A. Accetta, M. Pucci, and G. Vitale, "MRAS Speed Observer for High Performance Linear Induction Motor Drives based on Linear Neural Networks," *Power Electronics, IEEE Transactions on,* vol. PP, pp. 1-1, 2012.

[92] Z. Da and L. Hui, "A Stochastic-Based FPGA Controller for an Induction Motor Drive With Integrated Neural Network Algorithms," *Industrial Electronics, IEEE Transactions on,* vol. 55, pp. 551-561, 2008.

[93] Y.-J. Niu and X.-Y. Wang, "A novel adaptive fuzzy sliding-mode controller for uncertain chaotic systems," *Nonlinear Dynamics,* pp. 1-9, 2012.

[94] T. Orlowska-Kowalska, M. Dybkowski, and K. Szabat, "Adaptive Sliding-Mode Neuro-Fuzzy Control of the Two-Mass Induction Motor Drive Without Mechanical Sensors," *Industrial Electronics, IEEE Transactions on,* vol. 57, pp. 553-564, 2010.

[95] N. Noroozi, M. Roopaei, and M. Z. Jahromi, "Adaptive fuzzy sliding mode control scheme for uncertain systems," *Communications in Nonlinear Science and Numerical Simulation,* vol. 14, pp. 3978-3992, 2009.

[96] N. Yagiz, Y. Hacioglu, and Y. Taskin, "Fuzzy sliding-mode control of active suspensions," *Industrial Electronics, IEEE Transactions on,* vol. 55, pp. 3883-3890, 2008.

[97] W. Rong-Jong, "Fuzzy Sliding-Mode Control Using Adaptive Tuning Technique," *Industrial Electronics, IEEE Transactions on,* vol. 54, pp. 586-594, 2007.

[98] R. Pupadubsin, N. Chayopitak, D. G. Taylor, N. Nulek, S. Kachapornkul, P. Jitkreeyarn, et al., "Adaptive Integral Sliding-Mode Position Control of a Coupled-Phase Linear Variable Reluctance Motor for High-Precision Applications," *Industry Applications, IEEE Transactions on,* vol. 48, pp. 1353-1363, 2012.

[99] M. Comanescu, "An Induction-Motor Speed Estimator Based on Integral Sliding-Mode Current Control," *Industrial Electronics, IEEE Transactions on,* vol. 56, pp. 3414-3423, 2009.

[100] Z. H. Salih, K. S. Gaeid, and A. Saghafinia, "Sliding Mode Control of Induction Motor with Vector Control in Field Weakening," *Modern Applied Science,* vol. 9, p. p276, 2015.

[101] A. Bartoszewicz and A. Nowacka-Leverton, "SMC without the reaching phase-the switching plane design for the third-order system," *Control Theory & Applications, IET,* vol. 1, pp. 1461-1470, 2007.

[102] R. Pupadubsin, N. Chayopitak, D. G. Taylor, N. Nulek, S. Kachapornkul, P. Jitkreeyarn, et al., "Adaptive Integral Sliding Mode Position Control of a Coupled-Phase Linear Variable Reluctance Motor for High Precision Applications," *Industry Applications, IEEE Transactions on,* vol. PP, pp. 1-1, 2012.

[103] N. B. Cheng, L. W. Guan, L. P. Wang, and J. Han, "Chattering Reduction of Sliding Mode Control by Adopting Nonlinear Saturation Function," *Advanced Materials Research,* vol. 143, pp. 53-61, 2011.

[104] D. Y. Chen, W. L. Zhao, X. Y. Ma, and R. F. Zhang, "No-chattering sliding mode control chaos in Hindmarsh-Rose neurons with uncertain parameters," *Computers & Mathematics with Applications,* vol. 61, pp. 3161-3171, 2011.

[105] M. Roopaei, M. Zolghadri, and S. Meshksar, "Enhanced adaptive fuzzy sliding mode control for uncertain nonlinear systems," *Communications in Nonlinear Science and Numerical Simulation,* vol. 14, pp. 3670-3681, 2009.

[106] R. Shahnazi, H. M. Shanechi, and N. Pariz, "Position control of induction and DC servomotors: a novel adaptive fuzzy PI sliding mode

control," *Energy Conversion, IEEE Transactions on,* vol. 23, pp. 138-147, 2008.

[107] Z. Jinhui, S. Peng, and X. Yuanqing, "Robust Adaptive Sliding-Mode Control for Fuzzy Systems With Mismatched Uncertainties," *Fuzzy Systems, IEEE Transactions on,* vol. 18, pp. 700-711, 2010.

[108] C. Lascu, I. Boldea, and F. Blaabjerg, "A Class of Speed-Sensorless Sliding-Mode Observers for High-Performance Induction Motor Drives," *Industrial Electronics, IEEE Transactions on,* vol. 56, pp. 3394-3403, 2009.

[109] T. Orowska-Kowalska, M. Kaminski, and K. Szabat, "Implementation of a Sliding-Mode Controller With an Integral Function and Fuzzy Gain Value for the Electrical Drive With an Elastic Joint," *Industrial Electronics, IEEE Transactions on,* vol. 57, pp. 1309-1317, 2010.

[110] S. Rao, M. Buss, and V. Utkin, "Simultaneous State and Parameter Estimation in Induction Motors Using First- and Second-Order Sliding Modes," *Industrial Electronics, IEEE Transactions on,* vol. 56, pp. 3369-3376, 2009.

[111] L. Viet Quoc, C. Han Ho, and J. Jin-Woo, "Fuzzy Sliding Mode Speed Controller for PM Synchronous Motors With a Load Torque Observer," *Power Electronics, IEEE Transactions on,* vol. 27, pp. 1530-1539, 2012.

[112] L. Faa-Jeng, C. Po-Huan, C. Chin-Sheng, and L. Yu-Sheng, "DSP-Based Cross-Coupled Synchronous Control for Dual Linear Motors via Intelligent Complementary Sliding Mode Control," *Industrial Electronics, IEEE Transactions on,* vol. 59, pp. 1061-1073, 2012.

[113] M. Suetake, I. N. da Silva, and A. Goedtel, "Embedded DSP-Based Compact Fuzzy System and Its Application for Induction-Motor<formula formulatype="inline"> <img src="/images/tex/18925.gif" alt="V/f"> </formula> Speed Control," *Industrial Electronics, IEEE Transactions on,* vol. 58, pp. 750-760, 2011.

[114] P. Liu, C. Y. Hung, C. S. Chiu, and K. Y. Lian, "Sensorless linear induction motor speed tracking using fuzzy observers," *Electric Power Applications, IET,* vol. 5, pp. 325-334, 2011.

[115] H. Lee, E. Kim, H. J. Kang, and M. Park, "A new sliding-mode control with fuzzy boundary layer," *Fuzzy Sets and Systems,* vol. 120, pp. 135-143, 2001.

[116] F. Cupertino, D. Naso, E. Mininno, and B. Turchiano, "Sliding-Mode Control With Double Boundary Layer for Robust Compensation of

Payload Mass and Friction in Linear Motors," *Industry Applications, IEEE Transactions on,* vol. 45, pp. 1688-1696, 2009.

[117] Y. K. Kim and G. J. Jeon, "Error reduction of sliding mode control using sigmoid-type nonlinear interpolation in the boundary layer," *International Journal of Control, and Systems,* vol. 2, pp. 523-529, 2004.

[118] T. Kuo-Yang, L. Tsu-Tian, and W. Chi-Hsu, "Design of a New Fuzzy Suction Controller Using Fuzzy Modeling for Nonlinear Boundary Layer," *Fuzzy Systems, IEEE Transactions on,* vol. 13, pp. 605-616, 2005.

[119] C. M. Liaw and F. J. Lin, "Position control with fuzzy adaptation for induction servomotor drive," *Electric Power Applications, IEE Proceedings -,* vol. 142, pp. 397-404, 1995.

[120] W. Rong-Jong and S. Kuo-Ho, "Adaptive enhanced fuzzy sliding-mode control for electrical servo drive," *Industrial Electronics, IEEE Transactions on,* vol. 53, pp. 569-580, 2006.

[121] L. Mao-Fu, N. Michio, and H. Guan-Chyun, "Application of Fuzzy Logic in the Phase-Locked Loop Speed Control of Induction Motor Drive," *Industrial Electronics, IEEE Transactions on,* vol. 43, pp. 630-639, 1996.

[122] B. K. Bose, "Power electronics and AC drives," 1986.

[123] W. Leonhard, *Control of electrical drives*: Springer Verlag, 2001.

[124] C. C. Kung and K. H. Su, "Adaptive fuzzy position control for electrical servodrive via total-sliding-mode technique," *Electric Power Applications, IEE Proceedings -,* vol. 152, pp. 1489-1502, 2005.

[125] A. Emrouznejad and M. Tavana, *Performance measurement with fuzzy data envelopment analysis*: Springer, 2014.

[126] A. Amindoust and A. Saghafinia, "Supplier evaluation using fuzzy inference systems," in *Supply Chain Management Under Fuzziness*, ed: Springer, 2014, pp. 3-19.

[127] A. Amindoust, S. Ahmed, and A. Saghafinia, "Supplier Selection and Quota Allocation Decisions Under Uncertainty: Review and Future Research Directions."

[128] A. Amindoust, S. Ahmed, and A. Saghafinia, "A taxonomy and review on supplier selection methods under uncertainty," *International Journal of Information Technology and Business Management,* vol. 7, pp. 33-43, 2012.

[129] A. Amindoust, A. Shamsuddin, and A. Saghafinia, "Using data envelopment analysis for green supplier selection in manufacturing

under vague environment," in *Advanced Materials Research*, 2013, pp. 1682-1685.

[130] A. Amindoust, S. Ahmed, A. Saghafinia, and A. Bahreininejad, "Sustainable supplier selection: A ranking model based on fuzzy inference system," *Applied Soft Computing,* vol. 12, pp. 1668-1677, 2012.

[131] S. Sivanandam, S. Sumathi, and S. Deepa, *Introduction to fuzzy logic using MATLAB*: Springer Verlag, 2007.

[132] J. J. E. Slotine and W. Li, *Applied nonlinear control* vol. 461: Prentice hall Englewood Cliffs, NJ, 1991.

[133] W. J. Wang and J. Y. Chen, "A new sliding mode position controller with adaptive load torque estimator for an induction motor," *Energy Conversion, IEEE Transactions on,* vol. 14, pp. 413-418, 1999.

Chapter 3

NOVEL FUZZY SLIDING-MODE CONTROL FOR CHATTERING-FREE AND ROBUST INDUCTION MOTOR DRIVE

ABSTRACT

The boundary layer approach is the most popular method to reduce the chattering phenomenon in sliding mode control (SMC) for uncertain nonlinear systems. This chapter applies novel fuzzy sliding-mode speed controller based on boundary layer theory for an indirect filed-oriented control (IFOC) of induction motor (IM) drive. A fuzzy inference system is assigned for reaching controller part of fuzzy sliding mode controller (FSMC) to eliminate the chattering phenomenon in spite of the small and large uncertainties in the system. The applied fuzzy system acts like saturation function technique in thin boundary layer near to sliding surface so that stability of the system is guaranteed. Also, the equivalent control part is estimated to avoid the computational burden by an averaging filter. On the other hand, the averaging filter assists to improve of tracking performance despite the possibility of large uncertainties in the system so that the stability of the system is guaranteed. The main advantages of the proposed free-chattering speed controller are robustness to parameters variation and external load disturbance. The simulation results are performed to verify the effectiveness of the proposed speed controller, and its advantages are shown in comparison with the FSMC system and the conventional SMC.

Keywords: boundary layer variable, fuzzy sliding mode controller, speed control, induction motor

3.1. PREAMBLE

Some characteristics are important in electrical drives such as control of speed and torque in a different operating situation to motion control. In addition, issues dealing with the control of ac motor drives could be handled with vector control [1, 2].

In indirect vector method, estimation of the slip speed with appropriate information of the rotor time constant decouples flux and torque. The accuracy of motor parameters play important role for accuracy of this model, particularly the rotor time constant [3, 4]. However, field-orientation detuning is appeared by changing in the rotor time constant which in turn causes degradation of the induction motor drive performance. To overcome these problems, variable-structure control (VSC), and in particular, SMC system [5], have been recently applied for electrical motor drive.

To control electric drive systems, the SMC system as a robust control technique is an effective way with many attractive features as follows.1) it is robust to parameter variations and model uncertainties and insensitive to external load disturbance; 2) It offers a fast dynamic response, good transient performance, and stable control system; 3) it can handle some nonlinear systems that are not stable by using linear controller; and 4) it presents an easy hardware/software implementation. However, due to discontinuous control design, it has some limitation in electrical drives and shows high-frequency oscillations as chattering characteristics. The chattering can excite high frequency unmodeled dynamic of the system and make various undesirable effects such as current harmonics and torque pulsation [6-8].

In recent years, the chattering issue has become the research focus of many scholars [9-11]. Generally, introducing a thin boundary layer around the sliding surface can solve the chattering problem by interpolating a continuous function inside the boundary layer of switching surface [12, 13]. However, the slope of the continuous function is a compromise between control performance and chattering elimination [14, 15]. Also, asymptotic stability is not guaranteed and may cause steady-state error [16]. To recover these problems, a fuzzy controller is utilized to cope with the saturation function in reaching control part of the control effort in the SMC system.

The fuzzy theory was introduced by Zadeh [17]. According the fuzzy theory characteristics, the fuzzy approach is applied as an alternative method to handle uncertainty in plant dynamic in the past three decades [18-20]. The salient advantages of the designed fuzzy controller on the basis of the SMC system are decreasing the number of fuzzy rules and relaxation of the uncertainty bound requirement [21, 22]. The proper design of the FSMC system can eliminate the chattering phenomenon and improve the tracking performance despite the small uncertainty in nonlinear system [23, 24]. However, the systems with the possibility of large uncertainty like induction motor; by applying the FSMC system is not able to improve the tracking performance while the chattering phenomenon is omitted completely. In other words, the FSMC system is not able to reject disturbance in spite of the possibility of large uncertainty in nonlinear systems.

This chapter, replaces the saturation function in reaching control part by the suitable design of fuzzy inference system based on boundary layer theory so that the switching control law acts like saturation function technique in thin boundary layer near to sliding surface. To avoid the computational burden and improve the tracking performance despite the possibility of large uncertainties in the system, the equivalent control part is estimated by an averaging filter so that stability of the system is guaranteed.

3.2. THE FSMC SYSTEM

A boundary layer neighboring the switching surface is defined as [12],

$$b(t) = \{e(t), \quad |S(e(t);t)| \leq \Psi,\} \tag{3.1}$$

In the equation above, ψ is boundary layer thickness. In this state, $u(t)$ is selected as before (2.50) for outside of $b(t)$ and $u_{eq}(t)$, as best continuous approximation of $u(t)$, is interpolated inside of $b(t)$. In fact, the sign function is replaced by saturation function. Then, reaching control part (2.49) changes as,

$$u_r(t) = -(\overline{A}h)^{-1} k(t) sat(S(t)/\psi)) \tag{3.2}$$

where

$$sat(\frac{S}{\psi}) = \begin{cases} \dfrac{S}{\psi} & if \quad |S| \leq |\psi| \\ sign(S) & if \quad othewise \end{cases}$$

By applying the reaching control above, the control performance is not suitable while the chattering may be existed despite the possibility of large uncertainties in the system. To tackle these problems considering advantages of the boundary layer approach, the saturation function is replaced by the fuzzy inference system which acts like saturation function with a nonlinear slope inside the thin boundary layer.

Triangular type inputs membership function(MF) with fuzzy sets negative(N), zero(Z), and positive(P) and triangular and trapezoidal type output MF with fuzzy sets negative big(NB), negative medium(NM), negative small(NS), zero(ZE), positive small(PS), positive medium(PM), positive big(PB) on the common interval [-1 1] have been defined for the FSMC as shown in Figure 3.1 and Figure 3.2 respectively. According to these Figures, the thickness of boundary layer can be changed by varying the range of the fuzzy sets "Z" and "ZE" on the interval [0 1] in the inputs and output membership functions, respectively. Since the proposed fuzzy system structure is based on the saturation function technique, the best thickness of boundary layer can be derived from a fixed boundary layer sliding- mode controller which selects saturation function as reaching control part of its effort control. Thus, boundary layer thickness can be adjusted into two steps as follows [25].

Step 1: Varying the slope of saturation function so that the best performance is achieved for the fixed boundary layer sliding- mode based speed controller.

Step 2: Varying the range of the fuzzy sets "Z" and "ZE" to settle the selected value of boundary layer in the first step for the proposed FSMC based speed controller.

According to these steps considering Figure 3.1 and Figure 3.2 the value of the boundary layer thickness (ψ) is obtained equal to 0.85 in this chapter. The structure of designed fuzzy controller is shown in Figure 3.3.

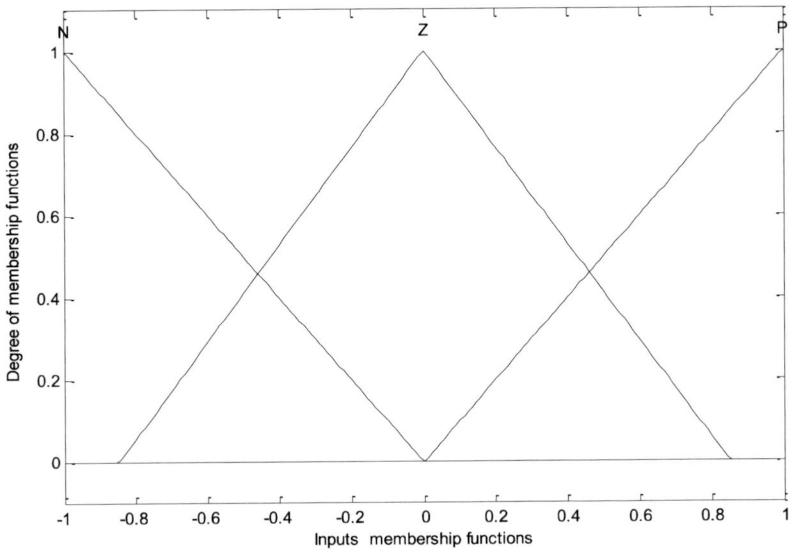

Figure 3.1. Inputs membership functions for the FSMC.

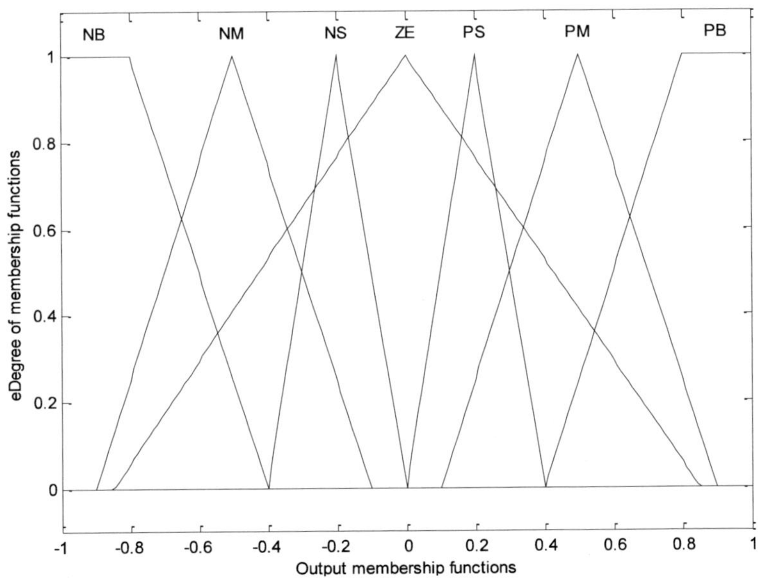

Figure 3.2. Output membership functions for the FSMC.

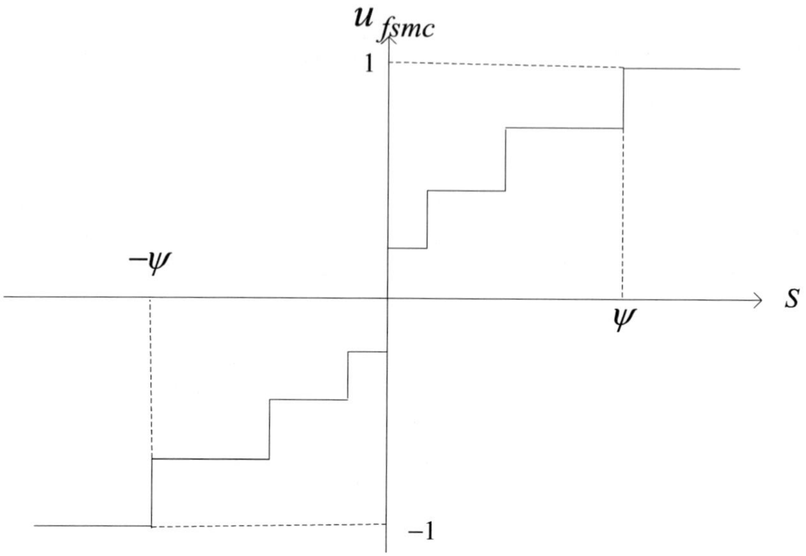

Figure 3.3. The performance of the FSMC system.

The FSMC system with two signals switching variable S , and alteration of switching variable as inputs and as output, maps input linguistic variable to output as,

$$u_{fsmc} = FSMC(S, \Delta S) \tag{3.3}$$

Then reaching control part changes as,

$$u_r(t) = -(\overline{A}h)^{-1} k(t) FSMC(S, \Delta S) \tag{3.4}$$

The Mamdani type fuzzy inferring method with 9 rules is applied to have appropriate tracking response, have fast dynamic response, eliminate chattering phenomenon, and satisfy the requirement of Stability condition (2.42) as shown in Table 1. The center of area (COA) method as defuzzification method is chosen for the FSMC system.

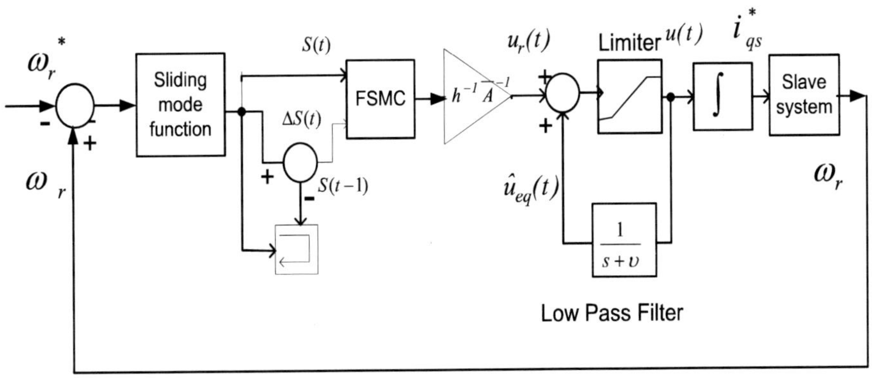

Figure 3.4. The control block diagram of the proposed model for IFOC induction motor.

Table 3.1. Fuzzy rules base with 9 rules for the FSMC system

ΔS \ S	N	Z	P
N	NB	NS	PM
Z	NB	ZE	PB
P	NM	PS	PB

The important considerations that have been taken into account for determining the rules are as follows [26].

1) When the switching variable S and the alteration of switching variable ΔS are far from the switching surface (whether positive or negative) then, the large control output is required to decrease of the quantity of S and ΔS. For instance, IF S is N and ΔS is N THEN u_{fsmc} is NB or IF S is P and ΔS is P THEN u_{fsmc} is PB.
2) When the switching variable S is outside of the boundary layer (see Figure 3.3) and it satisfies the stability condition (2.42) then, the control output should come back it to reach the switching surface in the same conditions. For instance, IF S is N and ΔS is P THEN u_{fsmc} is NM or IF S is P and ΔS is N THEN u_{fsmc} is PM.
3) When the switching variable S is inside of the boundary layer (see Figure 3.3) and it satisfies the stability condition (2.42) then, the

control output does not need to force it for reaching the switching surface quickly. For instance, IF S is Z and ΔS is P THEN u_{fsmc} is PS or IF S Z is and ΔS is N THEN u_{fsmc} is NS.

4) When the switching variable S and the alteration of switching variable ΔS are equal to zero then, the control output should be equal to zero.

In this stage, by applying the FSMC system along with the designed fuzzy rules and the mentioned conditions, the stability condition (2.42) is satisfied and consequently stability of system is guaranteed.

3.3. The Proposed Novel Fuzzy Sliding-Mode Speed Controller

Since the equivalent control makes derivative sliding function equal to zero, it can be obtained from (2.47). The equivalent control is actually the average of total control that can be estimated by an averaging filter. Essentially, the low frequency of a signal determines the characteristics of the signal and the high frequency is constructed due to unmodeled dynamic and uncertainties parameters of model. That is why; the equivalent control is estimated by a low pass filter which causes decreasing the computational burden [27]. On other hand, the proper break frequency selection of the low pass filter assists to improve the tracking performance despite the possibility of large uncertainties in the system. Then, the equivalent control is obtained as

$$\dot{\hat{u}}_{eq}(t) + \upsilon \hat{u}_{eq}(t) = u(t) \tag{3.5}$$

In equation above, υ is break frequency and it is selected by the trial and error method to omit the tracking error completely.

In this stage, the control law is obtained as,

$$u(t) = \hat{u}_{eq}(t) - (\overline{A}h)^{-1} k(t) FSMC(S, \Delta S) \tag{3.6}$$

The control block diagram of the proposed model for IFOC induction motor is shown as Figure 3.4. As shown in Figure 3.4, the controller in (3.6) must be passed through a limiter depend on limitations in inputs of plant.

3.4. Simulation Results

The simulation results are provided by using "Matlab" package due to periodic sinusoidal commands to verify the effectiveness of the proposed sliding-mode speed controller (the readers are referred to chapter 6 and [28] for more details). The simulation results show the advantages of the proposed speed controller in comparison with the FSMC system and the conventional SMC system. The simulation results by changing speed controller structure are obtained and shown in three stages as follows.

First Stage (conventional SMC system): choosing the conventional SMC system.

Second Stage: choosing the FSMC system.

Third Stage (the proposed speed controller): choosing the FSMC system along with the low pass filter as estimation for the equivalent control part.

The parameters of control in the each stage are adjusted so that the restriction of the control effort, the requirement of system stability, and the best transient control performance are provided. Appendix-A shows parameters of induction motor. In each stage of simulation, results due to periodic sinusoidal command with different cases including nominal situation, small uncertainty, and large uncertainty are addressed respectively as follows.

Case 1(nominal situation): using nominal parameter and external disturbance load T_L equal with zero.

Case 2(small uncertainty): inertia J_r and friction factor B are chosen two times of nominal value and external disturbance load T_L equal with 0.5Nm applied at 17sec.

Case 3 (large uncertainty): inertia J_r and friction factor is chosen two times of nominal value and external disturbance load T_L equal with 3Nm applied at 17sec.

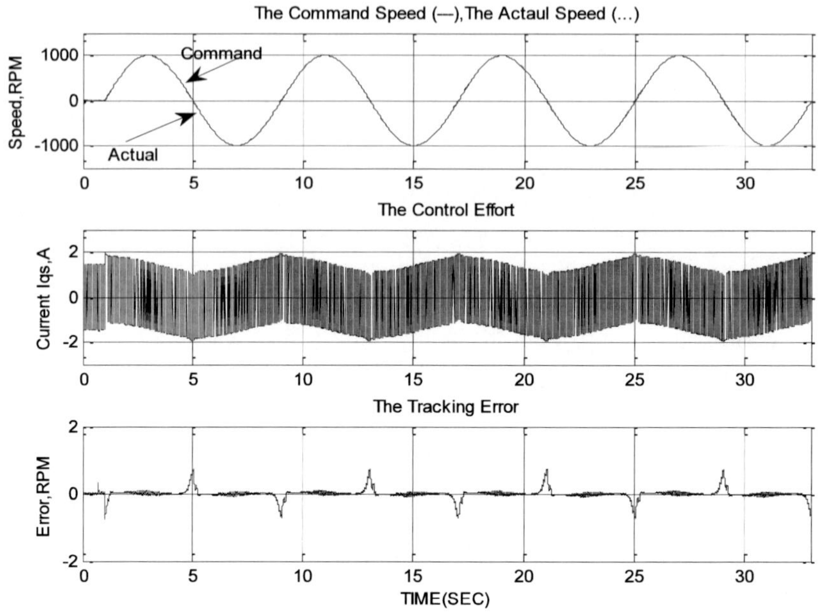

Figure 3.5. The conventional SMC system, the first stage in case 1, $k = 3$.

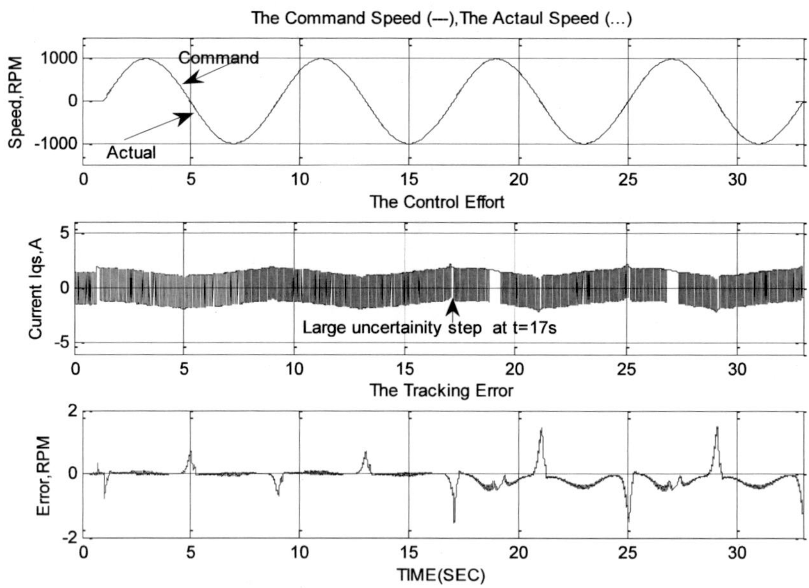

Figure 3.6. The conventional SMC system, the first stage in case 3, $k = 1$.

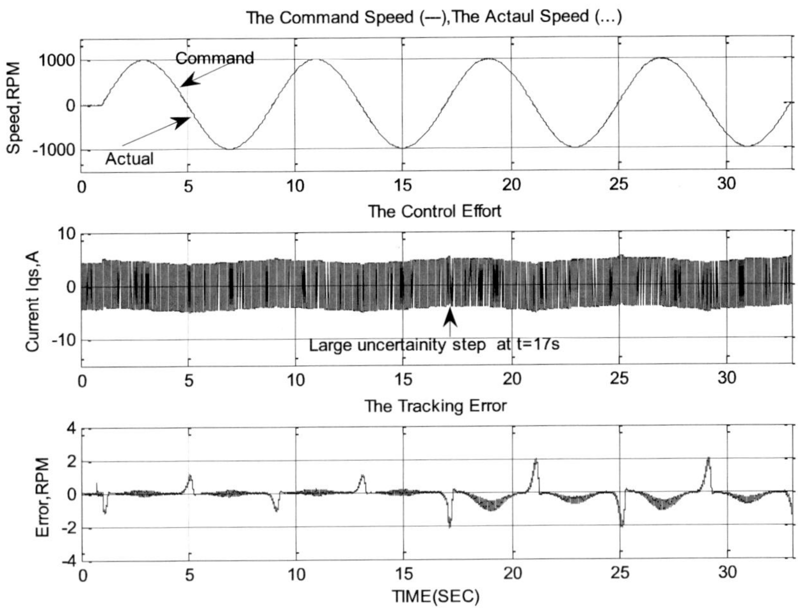

Figure 3.7. The conventional SMC system, the first stage in case 3, k = 3.

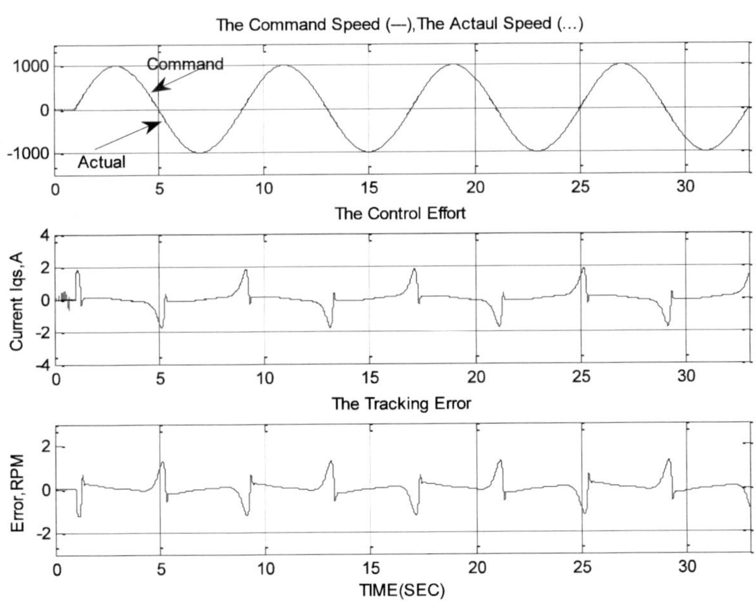

Figure 3.8. The FSMC system, the second stage in case 1, k = 15.

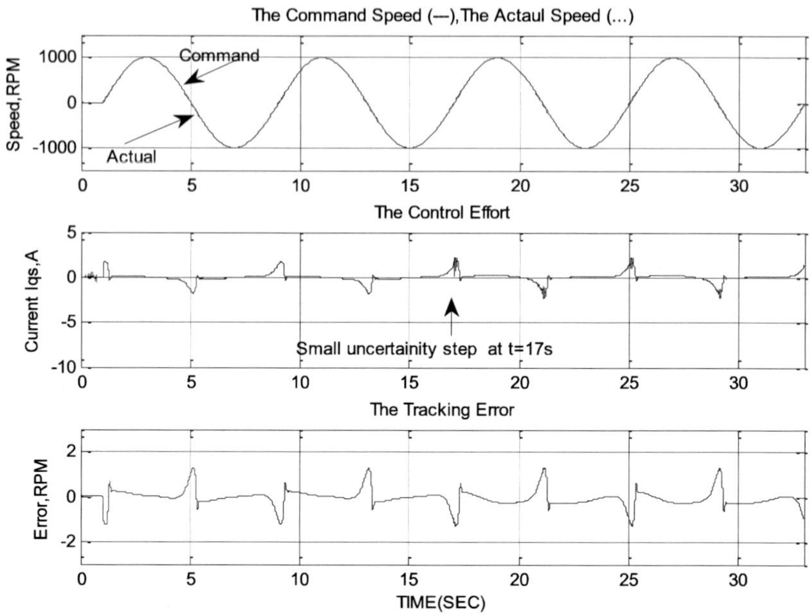

Figurre 3.9. The FSMC system, the second stage in case 2, $k = 15$.

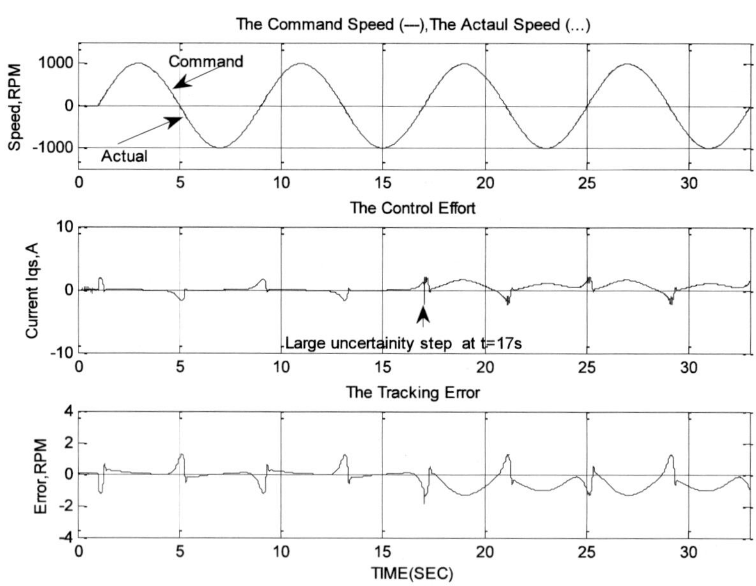

Figure 3.10. The FSMC system, the second stage in case 3, $k = 15$.

In the first stage, by choosing $C = 60, h = 1/377, \tau = 1$ and $k = 1$ the simulation result for conventional sliding mode in case 1 is illustrated as Figure 3.7. Figure 3.7 shows the sliding-mode controller has a favorable tracking performance with acceptable tracking error at the expense of undesirable chattering. However with utilization of the conventional SMC system in case 3, it is not able to handle unexpected and sudden unknown external disturbance as shown in Figure 3.6 and Figure 3.7 shows by increasing value of k to 3 in case 3, the conventional SMC system is able to cope with unexpected and sudden unknown external disturbance at the expense of excessive chattering that concludes degenerate tracking accuracy.

In the second stage, by choosing $C = 60, h = 1/377, \tau = 1$, and simulation result is obtained using the FSMC system. Figure 3.8 shows that the speed controller has the favorable tracking performance while the chattering is absent in case 1. By applying conditions in case 2, despite applying small sudden uncertainty, the chattering is still absent and the acceptable tracking error can be seen as shown in Figure 3.9. However, the FSMC system shows the large tracking error in control output in case 3. Figure 3.10 shows that it is unable to improve tracking performance in spite of the large uncertainty in the IFOC induction motor drive.

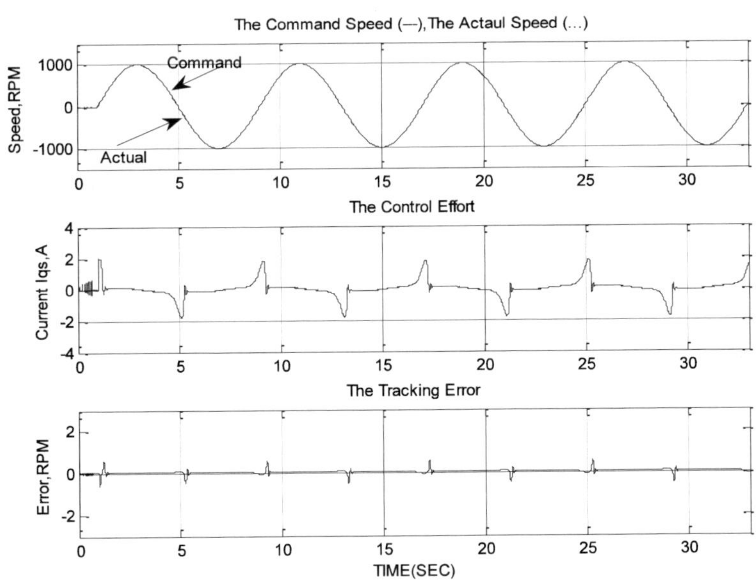

Figure 3.11. The proposed controller, the third stage in case1, $k = 15$.

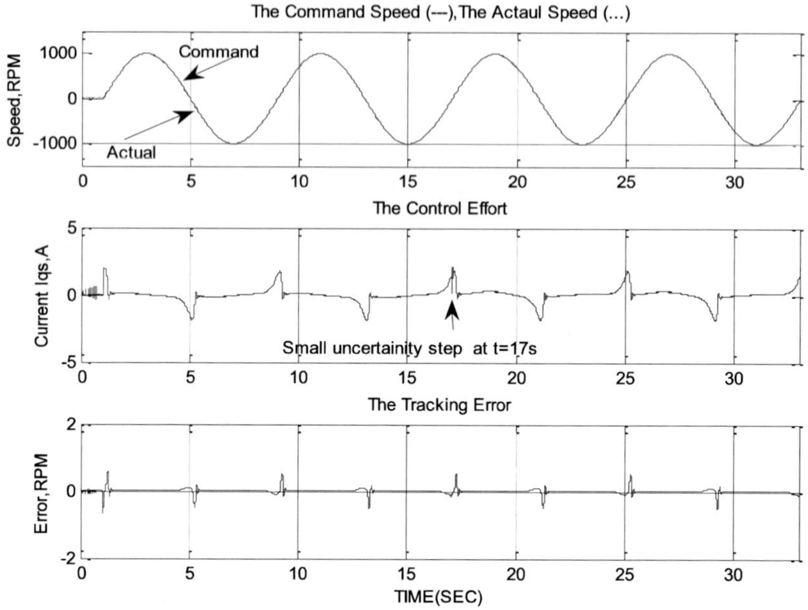

Figure 3.12. The proposed controller, the third stage in case 2, $k = 15$.

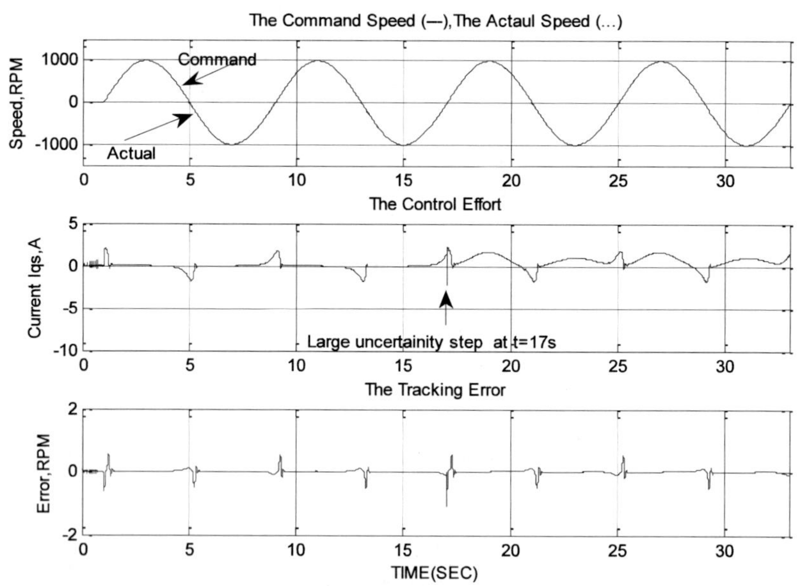

Figure 3.13. The proposed controller, the third stage in case 3, $k = 15$.

In the third stage, the proposed speed controller is implemented by choosing $C=60, h=1/377, \tau=1$, the break frequency $\upsilon=55$ and $k=15$. The simulation results are shown in Figure 3.11, Figure 3.12, and Figure 3.13. These Figures show that the proposed speed controller has the favorable tracking performance while the chattering has been omitted despite the small and large unexpected sudden unknown uncertainty. Simulation results indicate that the final model is robust to parameters variation and external load disturbance.

CONCLUSION

This chapter contributes to the presentation of a fuzzy sliding-mode structure based on boundary layer theory which is applied for speed controller of an IFOC induction motor drive. The used FSMC system is able to improve tracking performance and eliminate the chattering phenomenon for small unexpected sudden uncertainty of an IFOC induction motor drive. Also, the proper design of the low pass filter as estimation for the equivalent control part assists to the FSMC system for holding the tracking error near to zero in spite of the large unexpected sudden uncertainty. The main advantages of the proposed chattering-free speed controller are robustness to parameters variation and external load disturbance. Simulation results confirm the above-mentioned claims for the presented controller in the IFOC induction motor drive.

REFERENCES

[1] A. K. Chattopadhyay, "Advances in vector control ofac motor drives—A review," *Sadhana,* vol. 22, pp. 797-820, 1997.

[2] A. Saghafinia, H. W. Ping, and M. A. Rahman, "High Performance Induction Motor Drive Using Hybrid Fuzzy-PI and PI Controllers: a Review," *International Review of Electrical Engineering-Iree,* vol. 5, pp. 2000-2012, Sep-Oct 2010.

[3] M. Moallem, B. Mirzaeian, O. A. Mohammed, and C. Lucas, "Multi-objective genetic-fuzzy optimal design of PI controller in the indirect field oriented control of an induction motor," *Ieee Transactions on Magnetics,* vol. 37, pp. 3608-3612, 2001.

[4] A. Saghafinia and H. W. Ping, "High performance induction motor drive using fuzzy self-tuning hybrid fuzzy controller," in *Power and*

Energy (PECon), 2010 IEEE International Conference on, 2010, pp. 468-473.

[5] A. Bartoszewicz and A. Nowacka-Leverton, "SMC without the reaching phase-the switching plane design for the third-order system," *Control Theory & Applications, IET,* vol. 1, pp. 1461-1470, 2007.

[6] R. Shahnazi, H. M. Shanechi, and N. Pariz, "Position control of induction and DC servomotors: a novel adaptive fuzzy PI sliding mode control," *Energy Conversion, IEEE Transactions on,* vol. 23, pp. 138-147, 2008.

[7] Z. Jinhui, S. Peng, and X. Yuanqing, "Robust Adaptive Sliding-Mode Control for Fuzzy Systems with Mismatched Uncertainties," *Fuzzy Systems, IEEE Transactions on,* vol. 18, pp. 700-711, 2010.

[8] A. Saghafina, H. W. Ping, M. N. Uddin, and K. S. Gaied, "Adaptive fuzzy sliding-mode control into chattering-free induction motor drive," in *Industry Applications Society Annual Meeting (IAS), 2012 IEEE*, 2012, pp. 1-8.

[9] T. Orowska-Kowalska, M. Kaminski, and K. Szabat, "Implementation of a Sliding-Mode Controller with an Integral Function and Fuzzy Gain Value for the Electrical Drive With an Elastic Joint," *Industrial Electronics, IEEE Transactions on,* vol. 57, pp. 1309-1317, 2010.

[10] N. B. Cheng, L. W. Guan, L. P. Wang, and J. Han, "Chattering Reduction of Sliding Mode Control by Adopting Nonlinear Saturation Function," *Advanced Materials Research,* vol. 143, pp. 53-61, 2011.

[11] Z. H. Salih, K. S. Gaeid, and A. Saghafinia, "Sliding Mode Control of Induction Motor with Vector Control in Field Weakening," *Modern Applied Science,* vol. 9, p. p276, 2015.

[12] J. J. E. Slotine and W. Li, *Applied nonlinear control* vol. 461: Prentice hall Englewood Cliffs, NJ, 1991.

[13] F. J. Chang, S. H. Twu, and S. Chang, "Tracking control of DC motors via an improved chattering alleviation control," *Industrial Electronics, IEEE Transactions on,* vol. 39, pp. 25-29, 1992.

[14] H. Lee, E. Kim, H. J. Kang, and M. Park, "A new sliding-mode control with fuzzy boundary layer," *Fuzzy sets and systems,* vol. 120, pp. 135-143, 2001.

[15] F. Cupertino, D. Naso, E. Mininno, and B. Turchiano, "Sliding-Mode Control With Double Boundary Layer for Robust Compensation of Payload Mass and Friction in Linear Motors," *Industry Applications, IEEE Transactions on,* vol. 45, pp. 1688-1696, 2009.

[16] Y. K. Kim and G. J. Jeon, "Error reduction of sliding mode control using sigmoid-type nonlinear interpolation in the boundary layer," *International Journal of Control, and Systems,* vol. 2, pp. 523-529, 2004.

[17] L. A. Zadeh, "Fuzzy logic," *Computer,* vol. 21, pp. 83-93, 1988.

[18] A. Saghafinia, H. Wooi Ping, and M. Rahman, "High Performance Induction Motor Drive using Hybrid Fuzzy-PI and PI Controllers: a Review," *International Review of Electrical Engineering-Iree,* vol. 5, pp. 2000-2012, october-september 2010.

[19] Y. J. Xue and S. Y. Yang, "Synchronization of generalized Henon map by using adaptive fuzzy controller," *Chaos, Solitons & Fractals,* vol. 17, pp. 717-722, 2003.

[20] A. Amindoust, S. Ahmed, A. Saghafinia, and A. Bahreininejad, "Sustainable supplier selection: A ranking model based on fuzzy inference system," *Applied Soft Computing,* vol. 12, pp. 1668-1677, 2012.

[21] W. Rong-Jong and S. Kuo-Ho, "Adaptive enhanced fuzzy sliding-mode control for electrical servo drive," *Industrial Electronics, IEEE Transactions on,* vol. 53, pp. 569-580, 2006.

[22] N. Yagiz, Y. Hacioglu, and Y. Taskin, "Fuzzy sliding-mode control of active suspensions," *Industrial Electronics, IEEE Transactions on,* vol. 55, pp. 3883-3890, 2008.

[23] H. T. Yau and C. L. Chen, "Chattering-free fuzzy sliding-mode control strategy for uncertain chaotic systems," *Chaos, Solitons & Fractals,* vol. 30, pp. 709-718, 2006.

[24] M. Roopaei, M. Zolghadri, and S. Meshksar, "Enhanced adaptive fuzzy sliding mode control for uncertain nonlinear systems," *Communications in Nonlinear Science and Numerical Simulation,* vol. 14, pp. 3670-3681, 2009.

[25] A. Saghafinia, H. W. Ping, and M. N. Uddin, "Fuzzy sliding mode control based on boundary layer theory for chattering-free and robust induction motor drive," *The International Journal of Advanced Manufacturing Technology,* vol. 71, pp. 57-68, 2014.

[26] A. Saghafinia, H. W. Ping, M. N. Uddin, and K. S. Gaeid, "Adaptive Fuzzy Sliding-Mode Control Into Chattering-Free IM Drive," *Industry Applications, IEEE Transactions on,* vol. 51, pp. 692-701, 2015.

[27] M. Ertugrul, A. Sabanovic, and K. Ohnishi, "A generalized approach for Lyapunov design of sliding mode controllers for motion control applications," pp. 407-412 vol. 1.

[28] A. Saghafinia, H. W. Ping, M. N. Uddin, and A. Amindoust, "Teaching of Simulation an Adjustable Speed Drive of Induction Motor Using MATLAB/Simulink in Advanced Electrical Machine Laboratory," *Procedia-Social and Behavioral Sciences,* vol. 103, pp. 912-921, 2013.

Chapter 4

NOVEL BOUNDARY LAYER FUZZY CONTROL INTO CHATTERING-FREE AND ROBUST INDUCTION MOTOR DRIVE

ABSTRACT

This chapter presents a novel boundary layer fuzzy controller (NBLFC) based on the boundary layer approach for an indirect field-oriented control (IFOC) of an induction motor (IM) drive. The boundary layer approach leads to a trade-off between control performances and chattering elimination. For the NBLFC, a fuzzy system is used to adjust the boundary layer thickness to improve the tracking performance and eliminate the chattering problem under small uncertainties. Also, to eliminate the chattering under the possibility of large uncertainties, the integral filter is proposed inside the variable boundary layer. In addition, the stability of the system is analyzed through the Lyapunov stability theorem. The proposed NBLFC based IM drive is implemented in real-time using digital signal processor (DSP) board TI TMS320F28335. The experimental and simulation results show the effectiveness of the proposed NBLFC based IM drive at different operating conditions.

Keywords: variable boundary layer approach, boundary layer fuzzy controller, integral filter, speed control, induction motor

4.1. PREAMBLE

Vector control techniques with sensors or sensorless are very common in induction motor control applications due to their traditional superiority in high-performance applications. With the invention of the vector control technique the AC motor became popular for variable speed drives and motion control [1]. In indirect vector control, flux and torque are decoupled under estimation of the slip speed with appropriate information about the rotor time constant. The accuracy of motor parameters, particularly, the rotor time constant plays an important role for the accuracy of the indirect vector method [2-4]. In order to cope with that, recently, variable-structure control (VSC), and in particular, sliding-mode control (SMC) systems [5-8], have been applied for electric motor drives.

The SMC-based drive system has many attractive features [9] such as: (1) it is robust to parameter variations and model uncertainties are insensitive to external load disturbances; (2) it offers a fast dynamic response, and stable control system; (3) it can handle some nonlinear systems that are not stable by using a linear controller; and, (4) it only requires an easy hardware/software implementation. However, due to discontinuous nature, it has some limitations in electrical drives and shows high-frequency oscillations as chattering characteristics. This chattering produces various undesirable effects such as current harmonics and torque pulsations [10, 11].

In recent years, the chattering issue has become the research focus of many scholars [12, 13]. Generally, introducing a thin boundary layer around the sliding surface can solve the chattering problem by interpolating a continuous function inside the boundary layer of the switching surface [14, 15]. However, the slope of the continuous function is a compromise between control performance and chattering elimination [16]. Also, asymptotic stability is not guaranteed and may cause a steady-state error [17].

To improve tracking performance considering the thin boundary layer near the sliding surface, the slope of the continuous function or boundary layer thickness is adjusted by the fuzzy inference system [18, 19], which is called hereafter the conventional boundary layer fuzzy controller (BLFC). However, the authors in these works did not test the performance of IM drives with large disturbances, when the controller gets saturated and the performance of the device degrades. The IM drive often faces the possibility of large uncertainties, including large external load disturbances and variations of critical motor parameters in real-time. For large disturbances, the controller needs a high gain of the reaching control part and a thicker boundary layer to eliminate the chattering effects. On the other hand, increasing the boundary layer thickness decreases the feedback system to a system without sliding mode [20]. Hence, the conventional BLFC controller is not able to completely eliminate the chattering while it improves the tracking performance of the systems with the possibility of large uncertainty.

This chapter applies a modified fuzzy controller to adjust the thickness of the boundary layer near the sliding surface for improving tracking performance under small uncertainties. Also, an integral filter is proposed in the variable thin boundary layer to eliminate the chattering despite large uncertainties so that the stability of the proposed NBLFC is guaranteed. The performance of the proposed NBLFC-based IM drive is tested in both simulations and experiments and also compared with the conventional BLFC and PI controller-based IM drives.

4.2. The Existing BLFC Controller

A variable boundary layer neighbouring the switching surface is defined as [15],

$$b(t) = \{e(t), \quad |S(e(t);t)| \le \Psi, \quad \Psi = \psi(t)\} \qquad for \qquad \psi(t) > 0 \quad (4.1)$$

In the equation above, $\psi(t)$ is boundary layer thickness and it is allowed to vary with time. In this state, the stability condition must be modified due to the boundary layer thickness changes so that the distance to the boundary layer always decreases. It is guaranteed that all the trajectories outside the boundary layer are attracted toward the boundary [15]. Then, the stability condition (2.42) changes as,

$$\dot{V}(t) \le -(\eta - \dot{\psi}(t))|S(t)| \qquad for \ |S(t)| \ge \Psi(t) \qquad (4.2)$$

In this state, $u(t)$ is selected as before (2.50) for outside of $b(t)$ and $u_{eq}(t)$, as a best continuous approximation of $u(t)$, is interpolated inside of $b(t)$. However, if $\dot{\psi}(t) > 0$ during expansion boundary layer, it may not achieve asymptotic stability in stability condition [19]. So, in order to assure asymptotic stability, control gain $K(t)$ is added with the quantity ($-\dot{\psi}(t)$). Then, reaching control part (3.49) changes as,

$$u_r(t) = -(\overline{A}h)^{-1}\overline{k}(t)sat(S(t)/\psi(t)) \tag{4.3}$$

where, $\overline{k}(t) = k(t) - \dot{\psi}(t)$ and

$$sat(S(t)/\psi(t)) = \begin{cases} \dfrac{S}{\psi} & if \quad |S| \le |\psi| \\ \operatorname{sgn}(S) & if \quad othewise \end{cases}$$

The boundary layer thickness has been changed using the fuzzy inference system [18, 21] to eliminate the chattering. In this book, the proposed method used [19] is modified so that alteration of switching variable ($|\Delta S(t)|$) is added for switching variable $|S(t)|$ as input to the fuzzy inference system (BLFC) for the best control of the thin boundary layer thickness. The fuzzy rules are designed to reduce the tracking error while the reduction of the control error result in increasing the thickness of the boundary layer. Then, the boundary layer thickness as a nonlinear function of variable switching $|S(t)|$ and $|\Delta S(t)|$ is obtained as,

$$\psi(t) = BLFC(|S(t)|, |\Delta S(t)|) \tag{4.4}$$

Substituting (4.4) in (4.3) and considering (2.49), the reaching control and the control effort of the BLFC are determined, respectively as follows.

$$u_r(t) = -(\overline{A}h)^{-1}\overline{k}(t)sat(\psi(t), S(t)) = -\overline{k}(t)(\overline{A}h)^{-1}sat(BLFC(|S|, |\Delta S|), S(t)) \tag{4.5}$$

$$u(t) = -\overline{A}^{-1}[(C+\overline{B})\dot{e}(t) + \overline{B}\dot{\omega}_r^*(t) - \ddot{\omega}_r^*(t)] - \overline{k}(t)(\overline{A}h)^{-1}sat(BLFC(|S|, |\Delta S|), S(t)) \tag{4.6}$$

The Mamdani type fuzzy inferring method with 36 rules is used (Table 4.1). The center of the area (COA) method is used as defuzzification method. Triangular type membership functions with fuzzy sets as zero (*Z*), small (*S*), medium (*M*), medium big (*MB*), large (*L*), and very large (*VL*) have been defined on the interval [0, 1] for inputs and output as shown in Figure 4.1 and

Figure 4.2. Figure 4.3 shows the results inferred by the fuzzy rules. In this stage, stability condition (4.2) is satisfied by choosing the control law (4.6) and control gain as (2.53) and consequently the stability of the system is guaranteed for $|S(t)| > \psi(t)$.

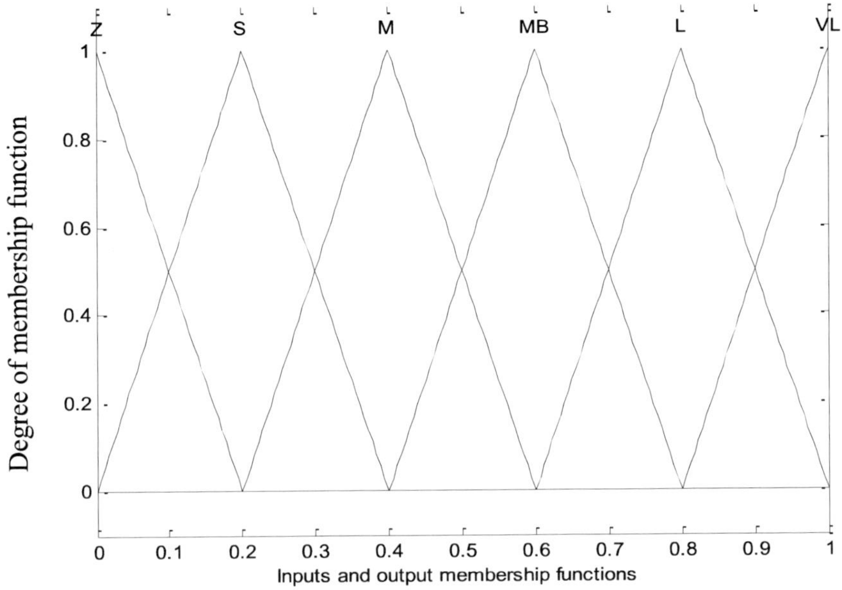

Figure 4.1. Inputs and output membership functions for the proposed NBLFC.

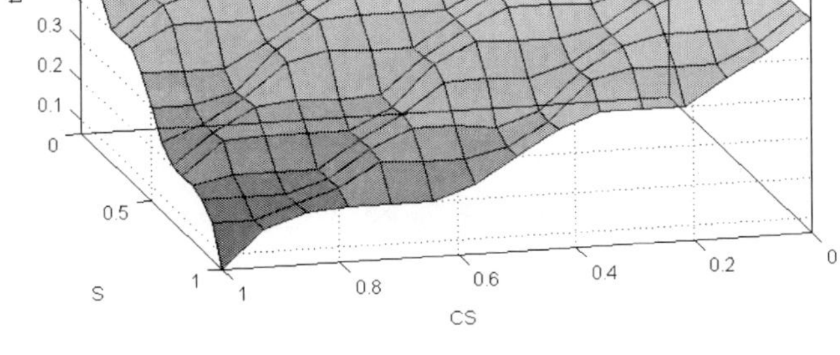

Figure 4.2. The 3D plot of input membership functions for the proposed NBLFC.

Table 4.1. Fuzzy rules base matrix for proposed NBLFC system

		Switching variable amplitude |s (t) |				
Alteration of switching variable amplitude |Δs(t)|		Z	S	M	MB	L	VL
	Z	VL	VL	L	L	MB	MB
	S	VL	L	L	MB	MB	M
	M	L	L	MB	MB	M	M
	MB	L	MB	MB	M	M	S
	L	MB	MB	M	M	S	S
	VL	MB	L	M	S	S	Z

4.3. The Proposed Novel Boundary Layer Fuzzy Controller

4.3.1. The Integral Filter Design

To eliminate the chattering under the possibility of large uncertainties from the existing BLFC, an integral filter is utilized inside the variable boundary layer, which discussed as follows [22].

For $|S(t)| \le \Psi(t)$ by substituting (4.6) in (2.46), yields:

$$\dot{S}(t)+\bar{k}(t)\frac{S(t)}{\psi(t)} = h\,\dot{L}(t) \tag{4.7}$$

Since $\bar{k}(t)$ and $\dot{L}(t)$ are continuous in ω_r, equation (4.7) can be written as below.

$$\dot{S}(t)+\bar{k}_d(t)\frac{S(t)}{\psi(t)} = h\,\dot{L}_d(t)+o(\zeta) \tag{4.8}$$

where $\bar{k}_d(t) = k_d(t) - \dot{\psi}(t)$ and $\dot{L}_d(t)$ is declared in desired state (ω^*_r) and $o(\xi)$ shows the error of replacing parameter $\omega_r(t)$ to $\omega^*_r(t)$. Variable switching $S(t)$ in (4.7) as output, whose dynamics only depends on the desired state $\omega^*_r(t)$, passes through a first- order low pass filter while lumped uncertainty is considered as input. Thus, the chattering can be eliminated or minimized.

The cut-off frequency of filter (4.7) is obtained as below.

$$\upsilon = \frac{\bar{k}_d(t)}{\psi(t)} \tag{4.9}$$

In the above equation, υ must be less than or equal c [23]. The value of the tracking precision ε equals to $\psi(t)/\upsilon$. Considering (4.9) and $\bar{k}_d(t) = k_d(t) - \dot{\psi}(t)$, the equation of $\psi(t)$ is obtained as,

$$\dot{\psi}(t) + \upsilon\psi(t) = k_d(t) \tag{4.10}$$

The variable switching $S(t)$ has a steady-state value depending on the inputs of the equation (4.8). So, increasing the control gain might increase $S(t)$ to achieve a precise tracking performance in spite of the possibility of large uncertainties in system. In this state, by using the thin boundary layer, it grows until it hits the boundary layer and when it is out of the boundary layer, it is imposed inward because of boundary layer attractiveness. So, the chattering is produced again. In fact, the discontinuity in sat function (4.5) at $S(t) = \psi$ causes the chattering [24]. To overcome this problem, the value of discontinuity is controlled and multiplied by a coefficient $\alpha(t)$ so that the coefficient $\alpha(t)$ fixes the bandwidth filter inside the boundary layer. So, the amount of discontinuity limits the value of the achievable filter bandwidth. To tackle the problem at this stage, an integral action with a coefficient $\beta(t)$ is added to force the trajectory on both sides of the boundary layer to face inward in order to prevent the chattering. Therefore the reaching control (4.5) changes as below.

$$u_r(t) = -\bar{k}(t)(\bar{A}h)^{-1}\, fsat(\psi(t), S(t), \alpha(t), \beta(t)) \tag{4.11}$$

where,

$$fsat(\psi(t), S(t), \alpha(t), \beta(t)) = \begin{cases} \alpha(t)\dfrac{S(t)}{\psi(t)} + \dfrac{\beta(t)}{\psi(t)}\int_0^t S(t)dt & if \quad |S(t)| \le |\psi(t)| \\ sign(S(t)) & if \quad othewise \end{cases}$$

In this stage, stability condition (4.2) is satisfied by choosing the reaching control (4.11) and the control gain (2.53) and consequently the stability of the system is guaranteed for $|S(t)| > \psi(t)$.

For $|S(t)| \le \psi(t)$ the system trajectory is declared in terms of the variable switching $S(t)$ as below.

$$\dot{S}(t) + \bar{k}_d(t)(\alpha(t)\frac{S(t)}{\psi(t)} + \frac{\beta(t)}{\psi(t)}\int_0^t S(t)dt) = h\dot{L}_d(t) + o(\zeta) \tag{4.12}$$

In the equation above, $\alpha(t)$ and $\beta(t)$ are chosen so that $S(t) \to 0$ for $t \to \infty$. Then, they can be chosen as follows.

$$\alpha(t) = \frac{2\upsilon\psi(t)}{\bar{k}_d(t)} \tag{4.13}$$

$$\beta(t) = \frac{\upsilon^2\psi(t)}{\bar{k}_d(t)} \tag{4.14}$$

In this state, the equation of $\psi(t)$ is obtained as,

$$\dot{\psi}(t) + 2\upsilon\psi(t) = k_d(t) \tag{4.15}$$

Substituting (4.13) and (4.14) in (4.12) and considering (4.9), equation (4.12) represents a filter as,

$$\dot{S}(t) + 2\upsilon(S(t) + \upsilon^2 \int_0^t S(t)dt) = h\dot{L}_d(t) + o(\zeta) \tag{4.16}$$

The polynomial roots of the filter above are obtained as below.

$$\frac{1}{p}(p^2 + 2\upsilon p + \upsilon^2) = 0 \tag{4.17}$$

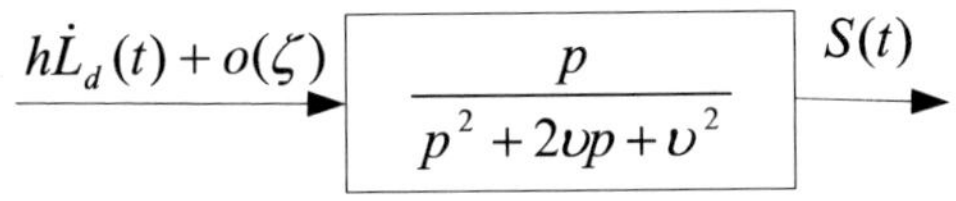

Figure 4.3. The integral filter inside boundary layer.

The lumped uncertainty is considered to be confined to input of the above filter so that $S(t)$ can be seen as its output as shown in Figure 4.3. Thus, choosing $\alpha(t)$, $\beta(t)$ as (4.13), (4.14) respectively, and the integral filter as (4.16), guarantees the stability of the system and the chattering phenomenon is eliminated completely. Therefore, the control law is obtained as

$$u(t) = u_{eq}(t) - \bar{k}(t)(\bar{A}h)\, fsat(\psi(t), S(t), \alpha(t), \beta(t)) \tag{4.18}$$

where,

$$fsat(\psi(t), S(t), \alpha(t), \beta(t)) = \begin{cases} \alpha(t)\dfrac{S(t)}{\psi(t)} + \dfrac{\beta(t)}{\psi(t)}\int_0^t S(t)dt & if \quad |S(t)| \le |\psi(t)| \\ sign(S(t)) & if \quad othewise \end{cases}$$

and

$$\psi(t) = BLFC(|S(t)|, |\Delta S(t)|)$$

The overall control block diagram of the new model for IFOC induction motor is shown as Figure 4.4. As shown in Figure 4.4, the controller in (4.18) must be passed through a limiter depending on the limitations inputs to the plant.

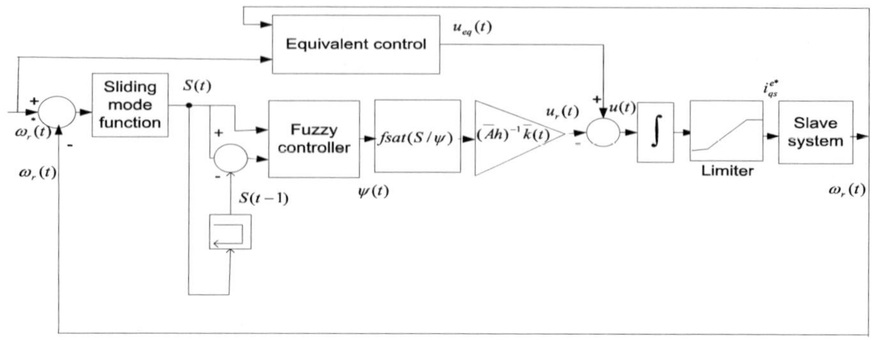

Figure 4.4. The control block diagram of the NBLFC based IM drive.

4.4. Simulation and Experimental Results

4.4.1. Simulation Results and Discussion

The performance of the proposed NBLFC-based IM drive has been investigated extensively in simulation. In order to show the superiority the performance of the proposed NBLFC is also compared with the tuned PI and the conventional BLFC controllers under different operating condition such as a sudden change of load, change of command speed, and parameter changes.

The discrete time Simulink model with sampling time Ts = 1×10^{-4} s along with the digital motor control (DMC) and IQMath libraries from TI and Mathworks are used to simulate the IFOC induction motor drive. These libraries were used to optimize the Simulink blocks. The SVM-VSI type inverter is modelled based on the fast switching IGBTs model from the Simulink toolbox along with the aforementioned libraries in MATLAB. Based on the block diagram of closed-loop vector control of the IM drive shown in Figure 2.1, a complete Simulink block diagram is designed, which is shown in Figure 4.5. TI and Mathworks provide a library of highly optimized and high precision math functions in the form of the IQMath library. IQMath is a way of representing a numeric value containing a sign, the integer and the fraction portion in a fixed bit location field. The IQ Math library is available in both fixed-and floating-point versions, enabling easy migration from float to fixed devices. These tools enable developers to quickly determine the processing resources required to implement basic motor control (the readers are referred to chapter 6 and [25] for more details). The PI controller is tuned by the

Ziegler–Nichols method based on the stability boundary [26, 27] to get the minimum overshoot/undershoot, minimum settling time and no steady-state error. The PI parameters are found as, Kp = 0.3, Ki = 0.0001, and Kc = 0.0001. The parameters of the IM are listed in Appendix-A.

For simulation tests, the following cases including parameter variations and external load disturbance are considered. If not mentioned, all other parameters are considered nominal in all the cases:

Case 1: Step changes in command speed with no load
Case 2: Step increase in load from '0' to 75% of rated load at t = 7 s.
Case 3: Inertia and friction coefficient are increased two times of nominal value at t = 7 s while half rated load is applied from the beginning.

Within the restriction of the control effort, in order to achieve the system stability, and the best transient performance, the control parameters are chosen as, $C = 1500, h = \overline{A}^{-1}, \tau = 1$, $K = 220$, and $\upsilon = 0.03$.

In **Case 1**, simulation results are illustrated in Figures 4.6-4.7, for the tuned PI, conventional BLFC and the proposed NBLFC controllers, respectively. From Figure 4.5, it can be seen that dynamic performance of the proposed NBLFC controller is almost similar to the conventional BLFC and better than those of the tuned PI controller. It can also be seen from Figures 4.6 that the conventional BLFC and the proposed NBLFC do not have any chattering in speed while maintaining almost the same tracking response.

Controllers, respectively, in **Case 2** are shown in Figures 4.8-4.9. It can be seen the PI controller suffers from a significant dip in speed (≈ 60 RPM) when the step increase in load is applied at t = 7 s. On the other hand, the conventional BLFC and the proposed NBLFC controllers are found to be almost insensitive when a step increase in load is applied. It is also found that the conventional BLFC suffers from chattering only when an external load disturbance is applied, whereas, it is found from Figures 4.8-4.9 that the proposed NBLFC is not only insensitive to load variations but also free from chattering in steady-state despite large uncertainty in the system. Moreover, the settling time for the proposed NBLFC based IM drive is faster than that of the tuned PI and almost similar to the conventional BLFC controller. Thus, the proposed NBLFC based IM drive ensures smooth operation of the motor and less harmonic losses in the motor.

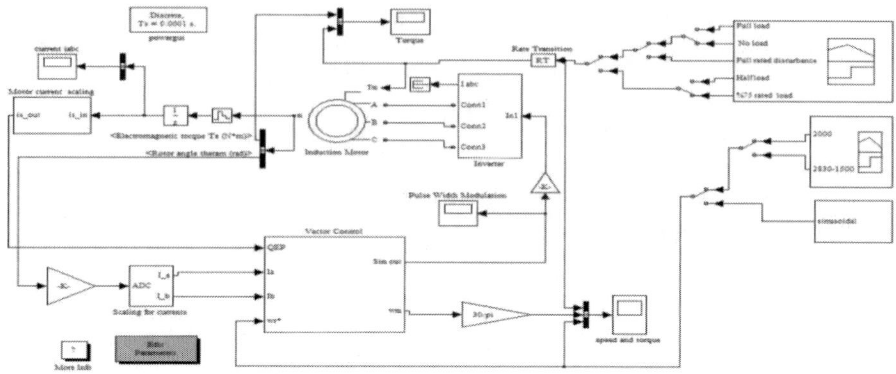

Figure 4.5. The main Simulink model for the IFOC of an IM drive.

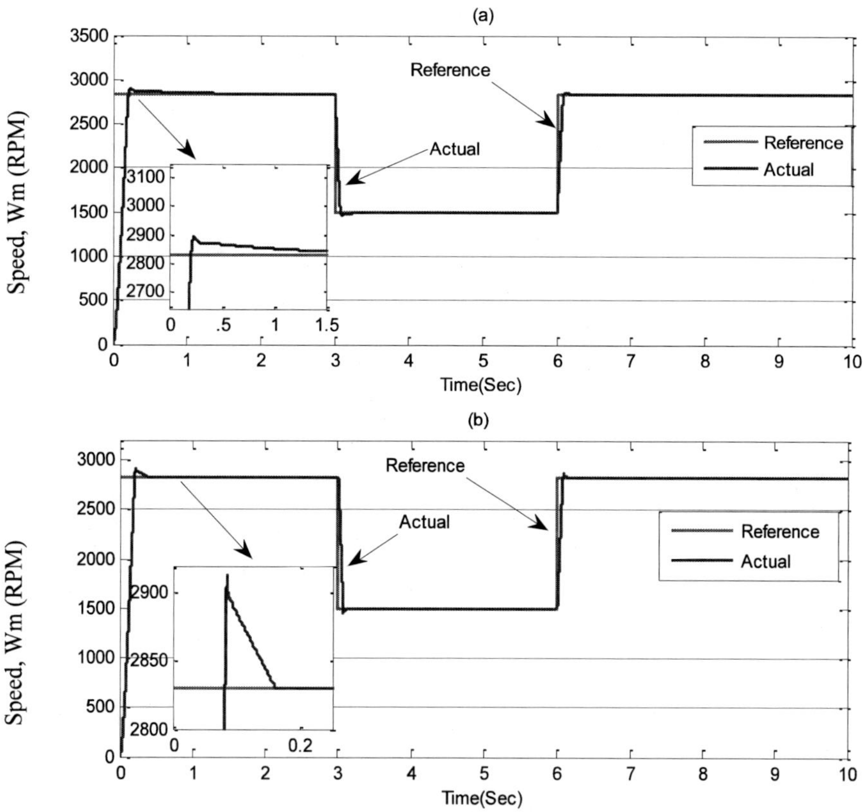

Figure 4.6. (Continued).

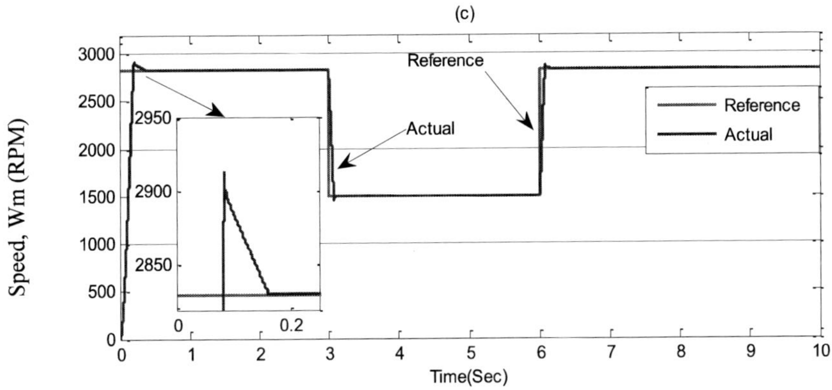

Figure 4.6. Simulated speed responses-based IM drive at no load in Case 1; (a) the tuned PI controller, (b) the conventional BLFC controller, and (c) the proposed NBLFC controller.

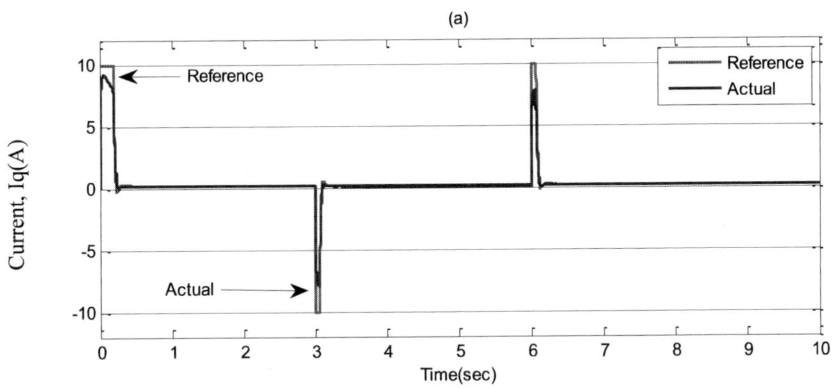

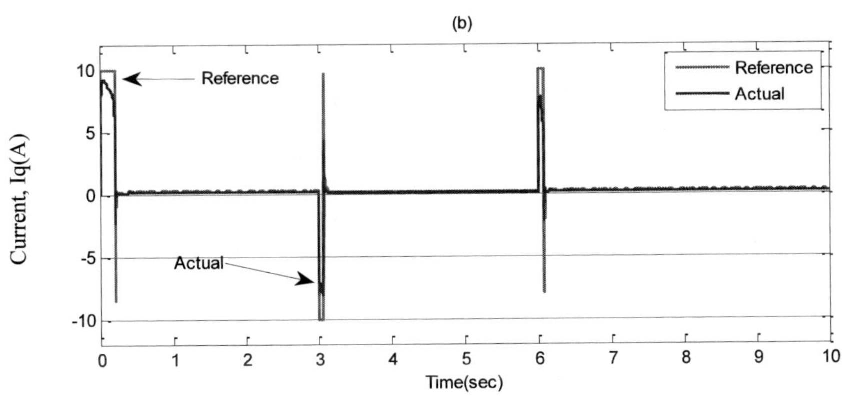

Figure 4.7. (Continued).

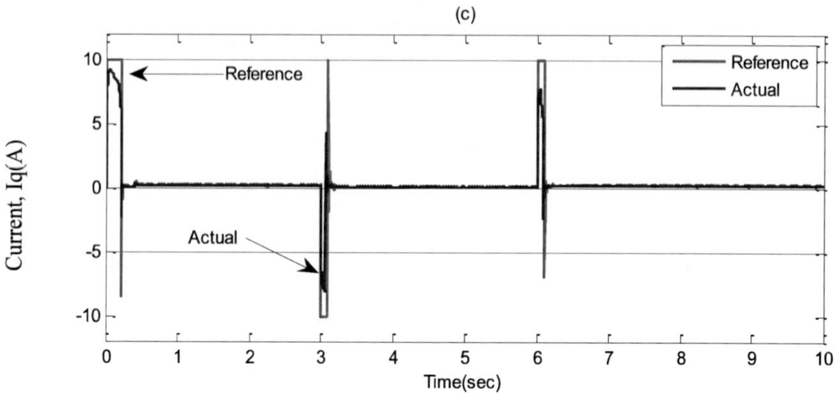

Figure 4.7. Simulated q-axis current responses-based IM drive at no load in Case 1; (a) the tuned PI controller, (b) the conventional BLFC controller, and (c) the proposed NBLFC controller.

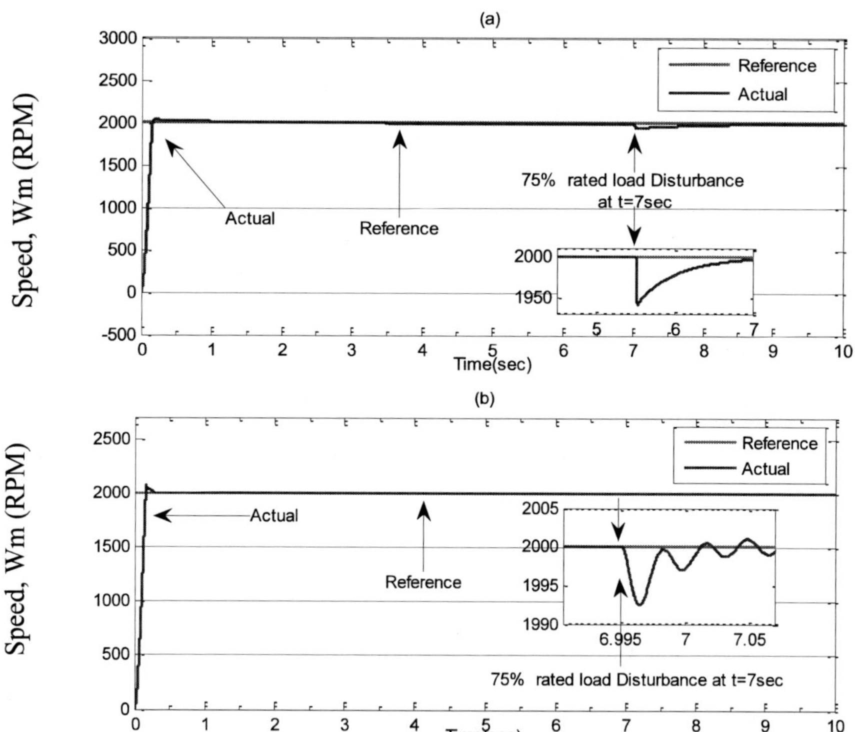

Figure 4.8. (Continued).

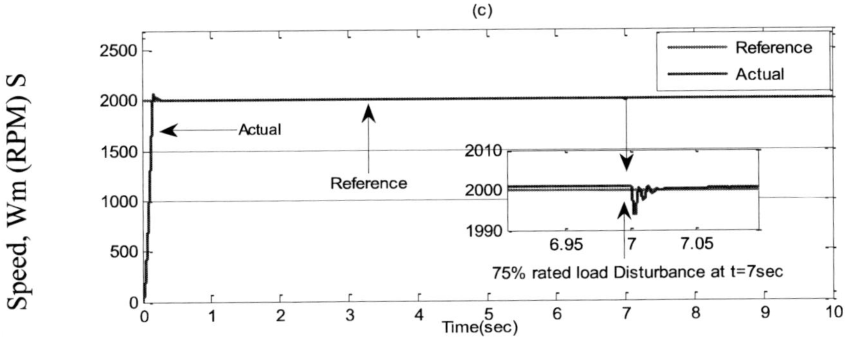

Figure 4.8. Simulated speed responses-based IM drive in Case 2; (a) the tuned PI controller, (b) the conventional BLFC controller, and (c) the proposed NBLFC controller.

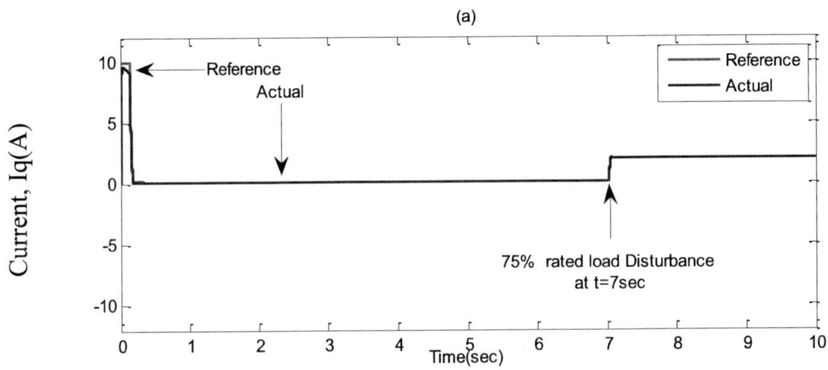

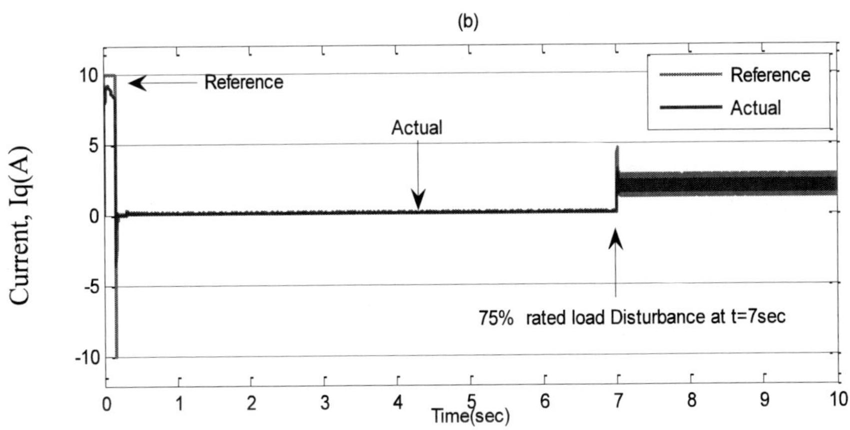

Figure 4.9. (Continued).

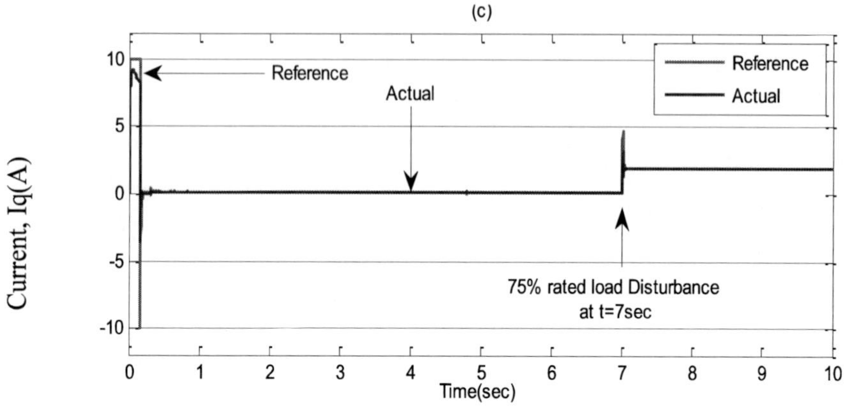

Figure 4.9. Simulated q-axis current responses-based IM drive in Case 2; (a) the tuned PI controller, (b) the conventional BLFC controller, and (c) the proposed NBLFC controller.

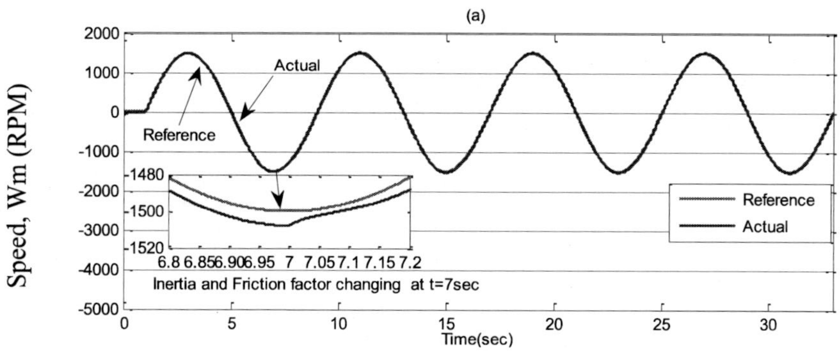

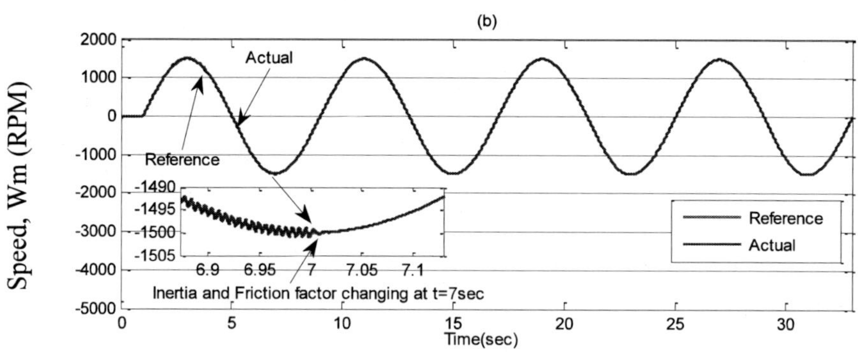

Figure 4.10. (Continued).

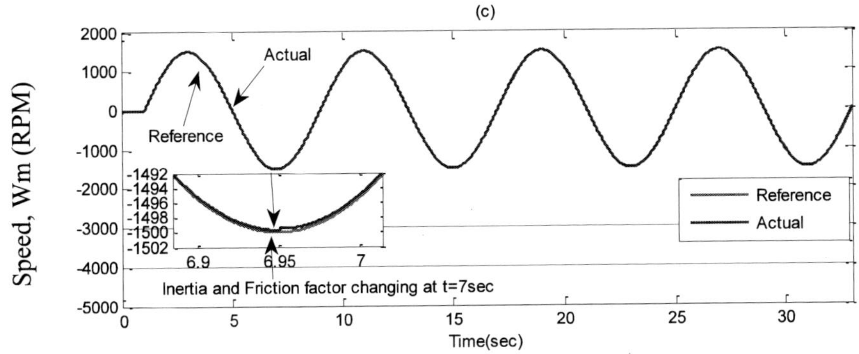

Figure 4.10. Simulated speed responses based IM drive in Case 3; (a) the tuned PI controller, (b) the conventional BLFC controller, and (c) the proposed NBLFC controller.

In **Case 3**, a sinusoidal command speed is selected to properly show the tracking error. The other parameter variations are also tested in this case. Simulation results are shown in Figures 4.10-4.11-4.12 for the tuned PI, the conventional BLFC, and the proposed NBLFC, respectively. It is found from these figures that the tracking capability of the conventional BLFC and proposed NBLFC is much better than the tuned PI controller. It is also found from these figures that the proposed NBLFC has acceptable tracking performance without any chattering, despite the large uncertainty in the system, while the conventional BLFC suffers from chattering for both speed and current. Due to the damping effect the conventional BLFC showed better response without chattering when the inertia and friction coefficient are increased. Thus, the highest accuracy in speed tracking of the proposed NBLFC with almost no chattering in current is verified in our simulation.

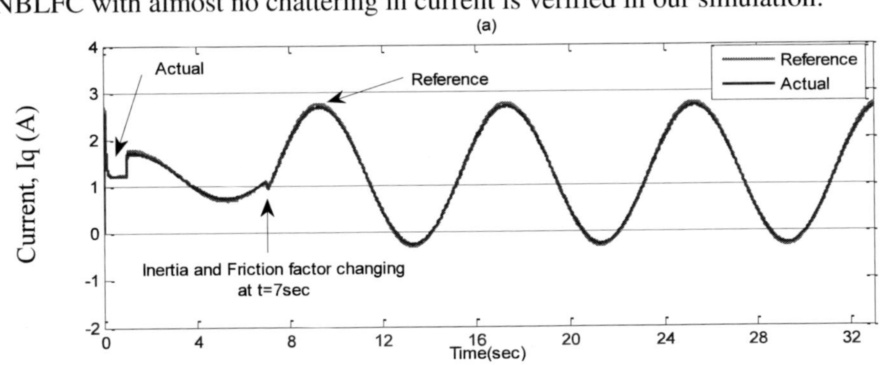

Figure 4.11 (Continued).

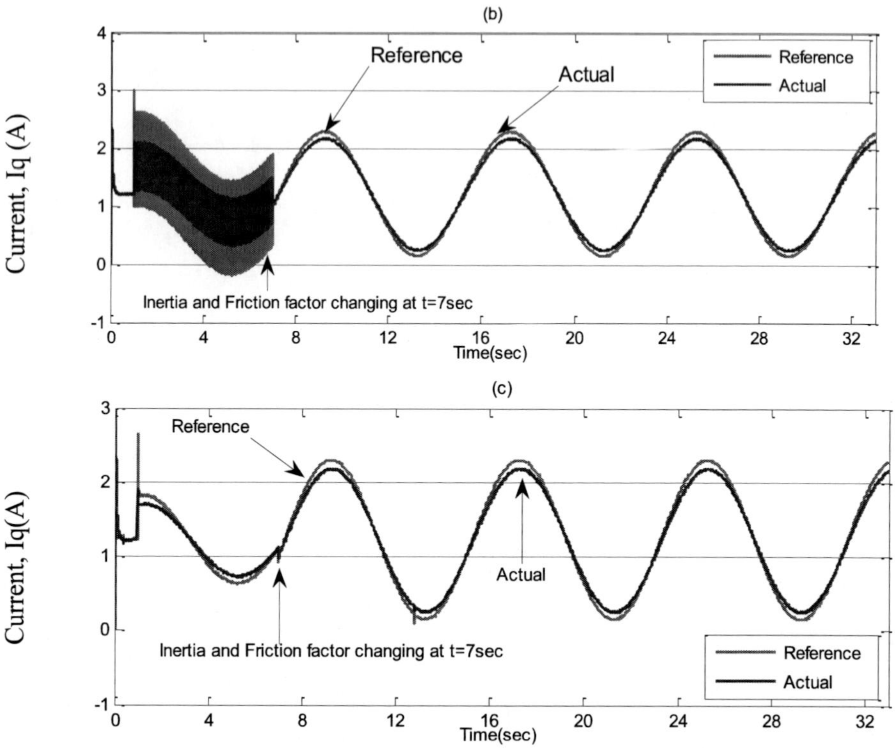

Figure 4.11. Simulated q-axis current responses-based IM drive in Case 3; (a) the tuned PI controller, (b) the conventional BLFC controller, and (c) the proposed NBLFC controller.

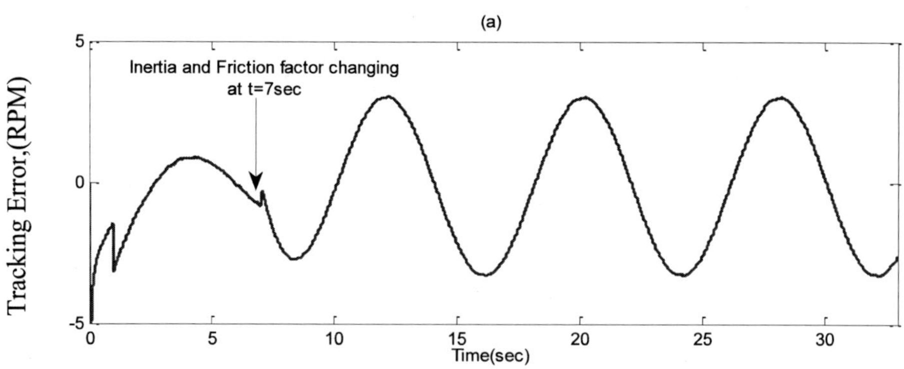

Figure 4.12. (Continued).

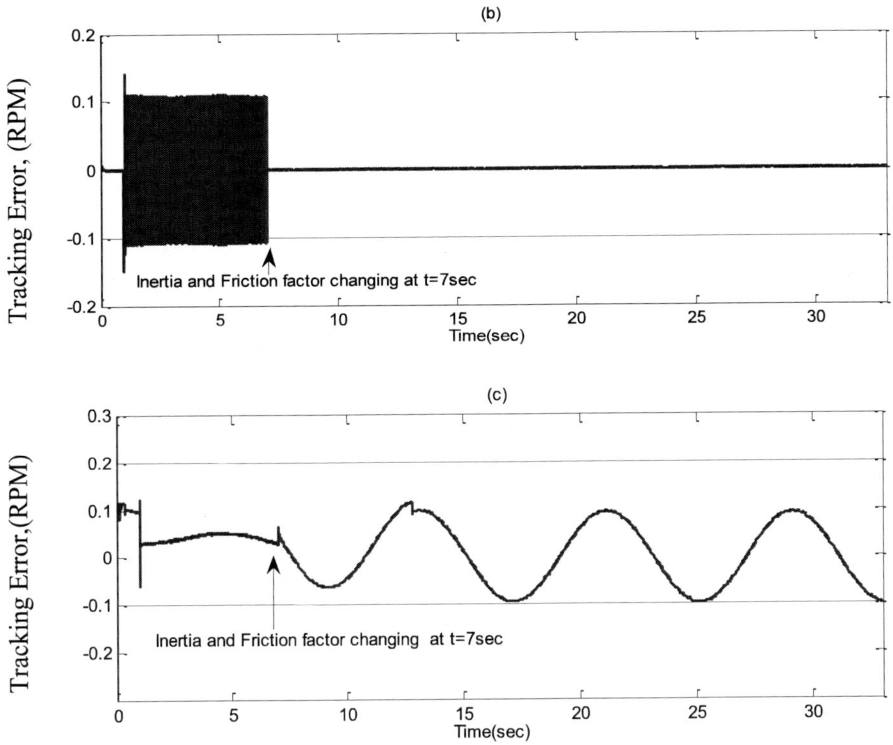

Figure 4.12. Simulated tracking error responses-based IM drive in Case 3; (a) the tuned PI controller, (b) the conventional BLFC controller, and (c) the proposed NBLFC controller.

4.4.2. Experimental Implementation

To implement the IM drive in real-time, an ezdspF28335 platform from Spectrum Digital is employed. Figure 4.13 shows a block diagram of the hardware schematic of a space vector modulated-voltage source inverter (SVM-VSI) fed IM drive. Three phase power inverters with 380V DC bus voltage, 10 kHz SVM-PWM switching frequency and 2 μs dead time for short circuit protection are used to drive the IM (the readers are referred to chapter 7 and for more details).

As shown in Figure 4.13, the phase currents are sensed by the two HX 10-P/SP2 current transducer sensors, which have a current range of ±10 A. The current signals are fed back to the ezdspF28335 board through analogue to digital converter (A/D) channels. The output current signal of these sensors is converted to a voltage across the resistor connected (2 KΩ, 1/4 W) between the output terminal of the sensor and ground. These currents are fed to the DSP board through the A/D converter. Due to the noises from current sensors and A/D of the digital signal processor (DSP), there are lots of spikes in the motor currents, which cannot meet the requirements of vector control in the case of small signals. To remove these noises, there are some potential solutions such as using a suitable voltage source for the ADC conditioning circuit, a proper electronic design layout, using a small ceramic capacitor along with the electrolytic capacitor, using the proper op-amps [28-32], and using digital filters to omit the noise from motor current signals. In this work considering all of the aforementioned notices, a digital low pass filter with a suitable cut-off frequency (normally chosen above the chattering frequency) is employed. It is found that the best cut-off frequency is about 500 Hz [22].

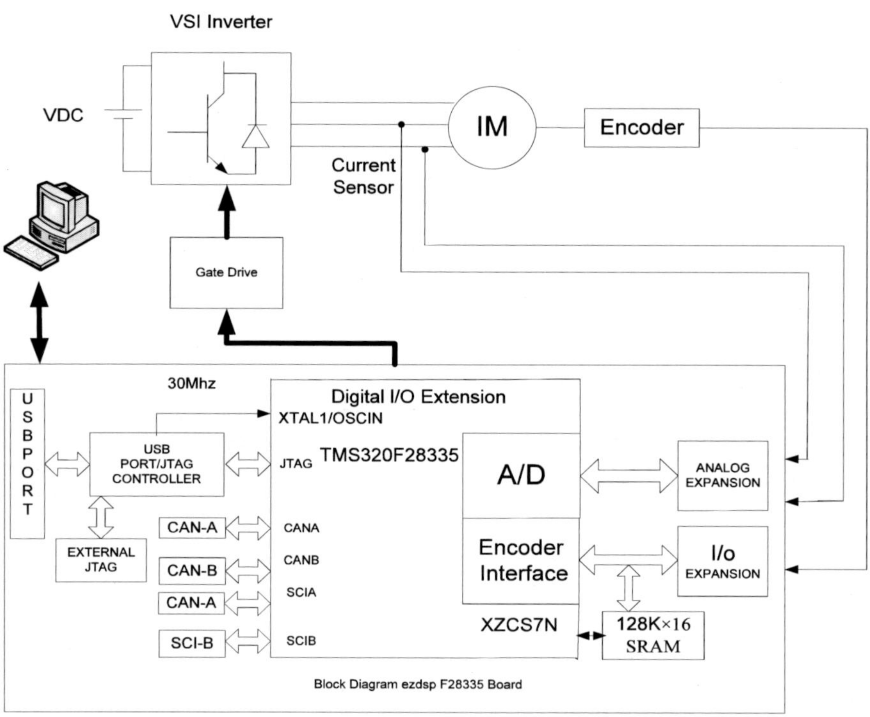

Figure 4.13. Block diagram of the hardware schematic for real-time implementation of VSI fed IM drive.

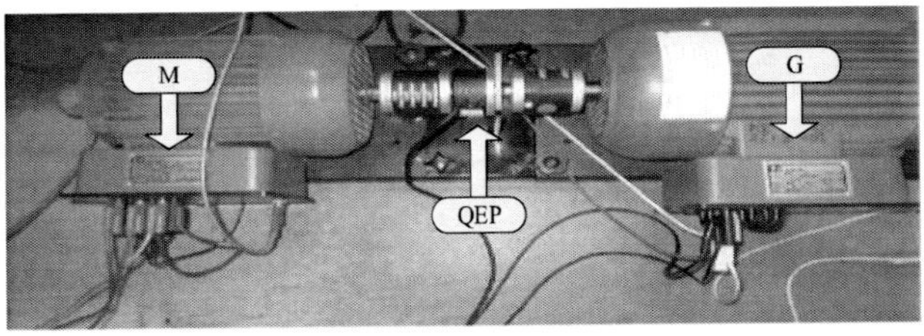

Figure 4.14. Connection of encoder for IM drive.

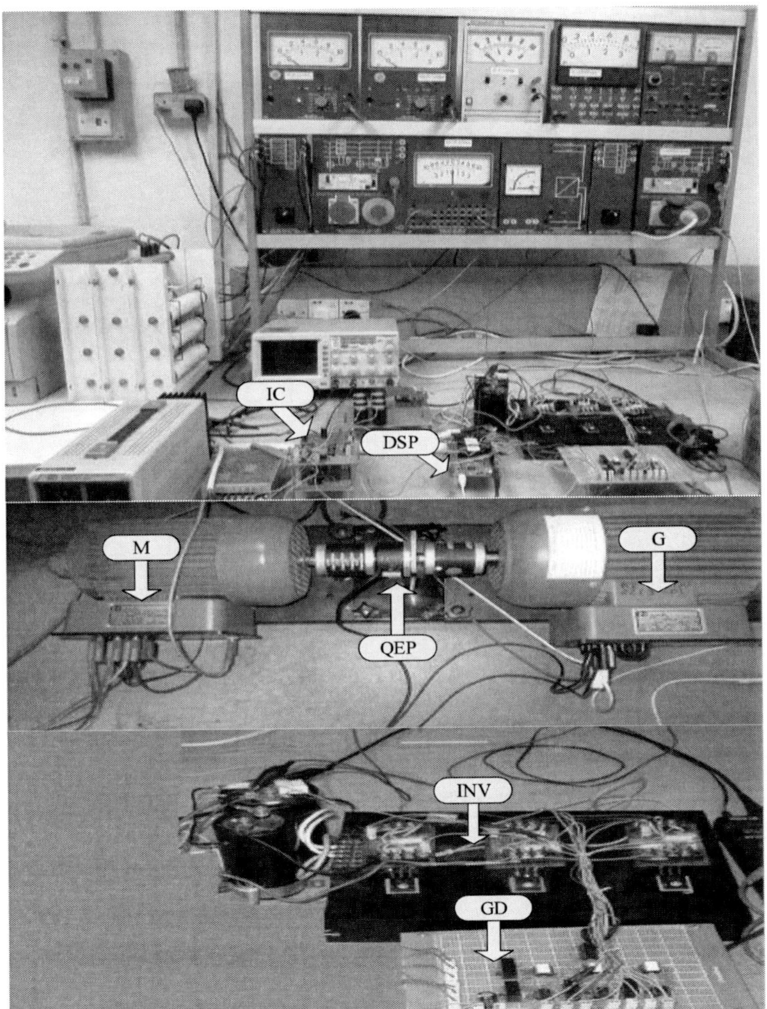

Figure 4.15. Experimental setup of the proposed IM drive.

Rotor position is sensed by an optical incremental encoder (QEP) E60H20 with three output channels (A, B, and index) which must be mounted properly on the rotor shaft as shown in Figure 4.14 and is fed back to the ezdspF28335 through the I/O expansion as seen in Figure 4.13. The used encoder generates 5000 pulses per turn which is multiplied by 4 using the quadrature mode, so the resolution of the encoder is $2\pi/(5000\times4)=0.018^\circ$. A 16-bit position counter is used to count the encoder and it is reset in each revolution by an index pulse. Since the encoder is an open collector type with 15 V power supply, a 1 KΩ pull-up resistance is connected between the power supply and each output and some resistive voltage dividers are used to reduce the output voltages to 3.3 V to meet the DSP voltage conditions. This optical incremental encoder also used to calculate high-resolution speed measurements. It is noted that all related requirements for connection the sensors to DSP are provided with an interface circuit (IC) as shown in Figure 4.15.

The control algorithms are made by Simulink models and downloaded to the DSP board through Code Composer Studio (CCStudio) TI software. The outputs of the board are six logic signals, which are fed to the inverter (INV) through the gate drive (GD) circuit as shown in Figure 4.15. Due to the computational burden of the fuzzy controller, the sampling frequency of the proposed controller should be less than the PI controller in real-time to get the results properly. Thus, the sampling frequencies of experimental implementation are set at 10 kHz and 4 kHz for the tuned PI and the proposed controllers, respectively. The necessary data is saved on DSP's memory with 400 Hz sampling frequency. A DC generator is coupled with the IM as a load.

The experimental setup for the proposed NBLFC based prototype 1 kW IM drive system is shown in Figure 4.15.

4.4.3. Experimental Results and Discussion

The tuned PI, the conventional BLFC, and the proposed NBLFC were simulated to search for the controller which provides the best dynamic performance. Through simulation tests it was found that the proposed NBLFC provides the best dynamic performance. To validate the simulation results, the proposed controller is implemented in real-time and compared with the conventional BLFC and the tuned PI controller based IM drive. As the PI controller has been utilized widely in the industry, it was considered as the benchmark controller. For the experimental test verifications, the following cases are considered. If not mentioned, like in the simulation results, all other parameters are considered nominal in all the cases:

Case 1: Step changes in command speed with no load.
Case 2: Step increase in load from '0' to full load.

It is noteworthy that the PI parameters in practice, are found as, Kp = 0.8, Ki = 0.0002, and Kc = 0.0002. In **Case 1**, experimental results are illustrated in Figures 4.16-4.17 for the tuned PI and the proposed NBLFC based IM drive, respectively. From Figure 17, it can be seen that speed response of the conventional BLFC controller and the proposed NBLFC are better than that of the tuned PI controller in terms of rising, overshoot, and settling times with the step changes in command speeds. In addition, From Figure 4.16, it can be seen that dynamic performance of the proposed NBLFC controller is better than that of the conventional BLFC controller in terms of rising time. It can be seen from Figure 4.17 that the proposed NBLFC controller exhibit significantly less steady-state chattering in current as compared to the conventional BLFC and the tuned PI.

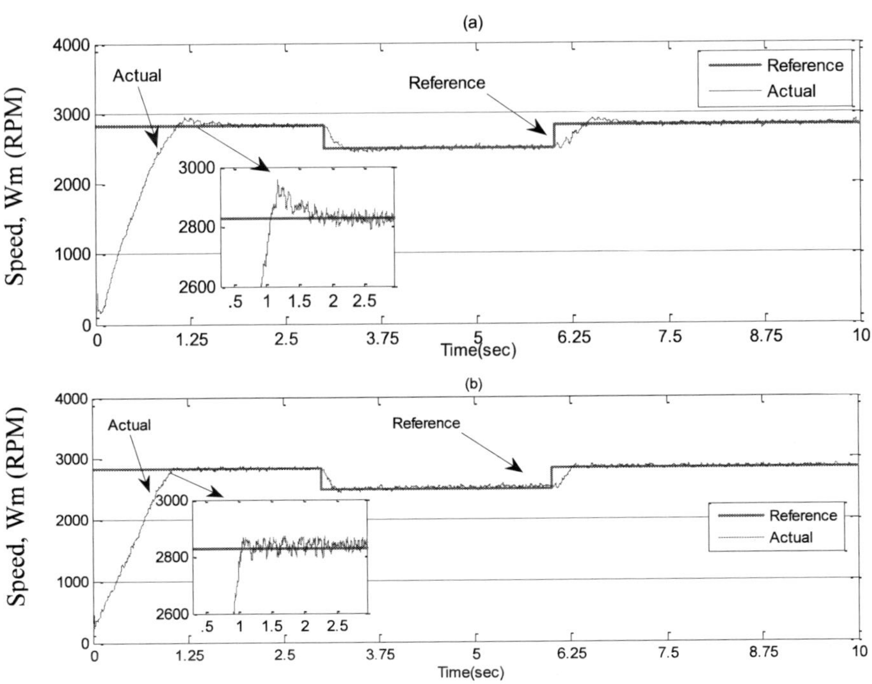

Figure 4.16. (Continued).

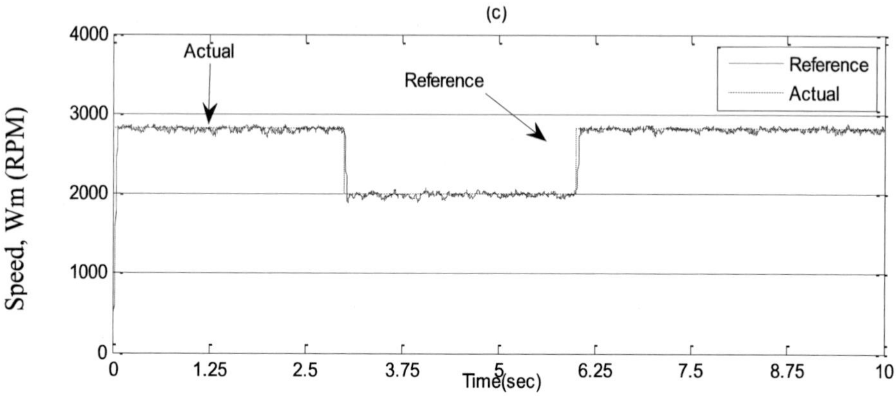

Figure 4.16. Experimental speed responses-based IM drive in Case 1; (a) the tuned PI controller, (b) the conventional BLFC controller, and (c) the proposed NBLFC controller.

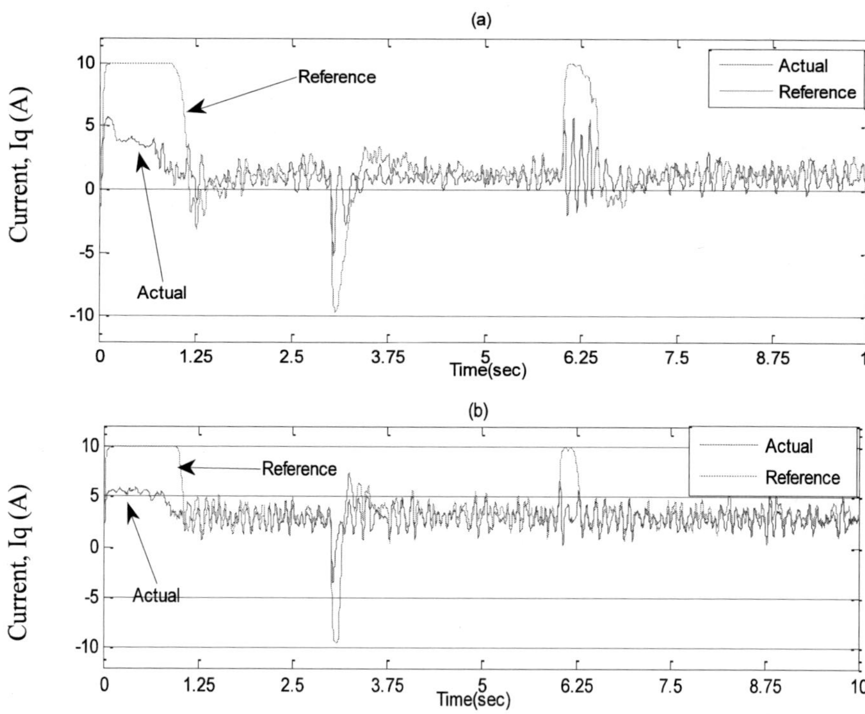

Figure 4.17. (Continued).

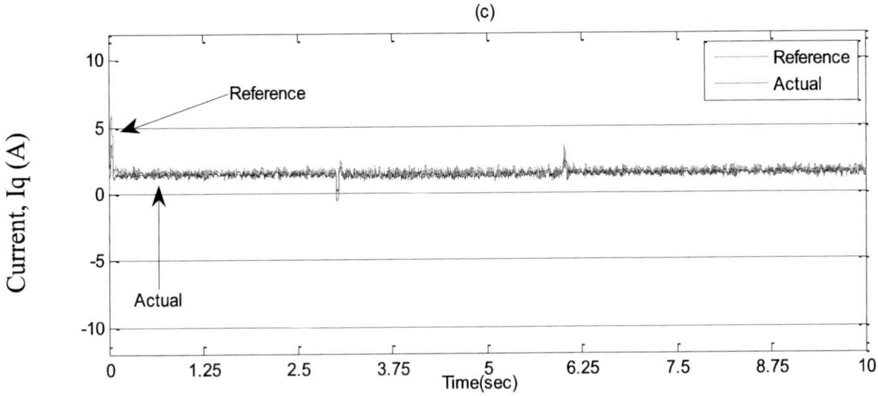

Figure 4.17. Experimental q-axis current responses based IM drive in Case 1; (a) the tuned PI controller, (b) the conventional BLFC controller, and (c) the proposed NBLFC controller.

In **Case 2**, experimental results are shown in Figures 4.18-4.19 for a step increase in full load. The motor was running under no load condition, then, suddenly a full load disturbance is applied to the motor. From these Figures, it can be seen the PI controller suffers from a significant dip in speed (≈ 100 RPM) when the step increase in load is applied. Moreover, the rising time for the proposed NBLFC based IM drive is faster than that of the conventional BLFC and the tuned PI controller. It is also found that the conventional BLFC suffers from chattering when the external load disturbance is applied, whereas, it is found from Figures 19-20 that the proposed NBLFC is not only insensitive to load variations, but also free from chattering in steady-state despite the large uncertainty in the system. Thus, the proposed NBLFC generates less harmonic loss and associated heat in the motor, which is a major issue in real-time.

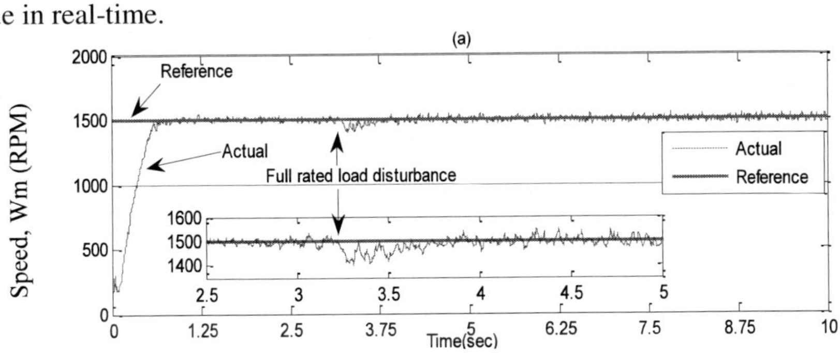

Figure 4.18. (Continued).

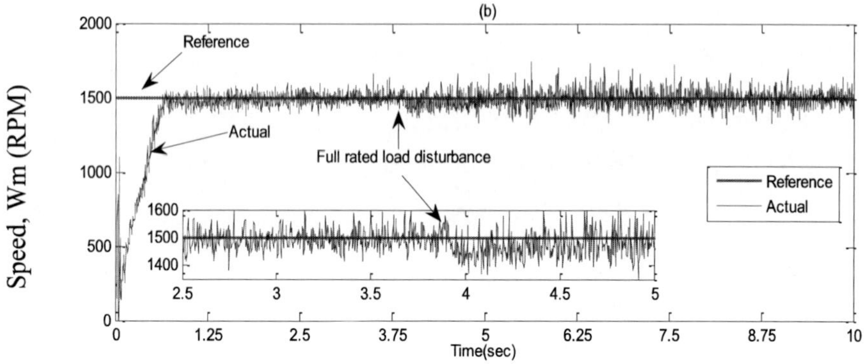

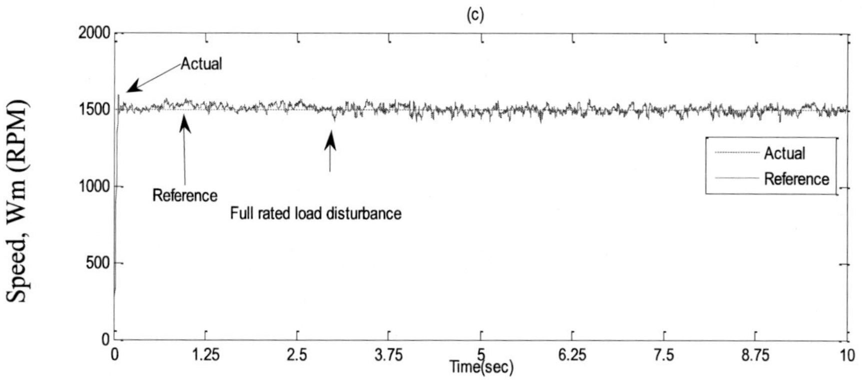

Figure 4.18. Experimental speed responses-based IM drive in Case 2; (a) the tuned PI controller, (b) the conventional BLFC controller, and (c) the proposed NBLFC controller.

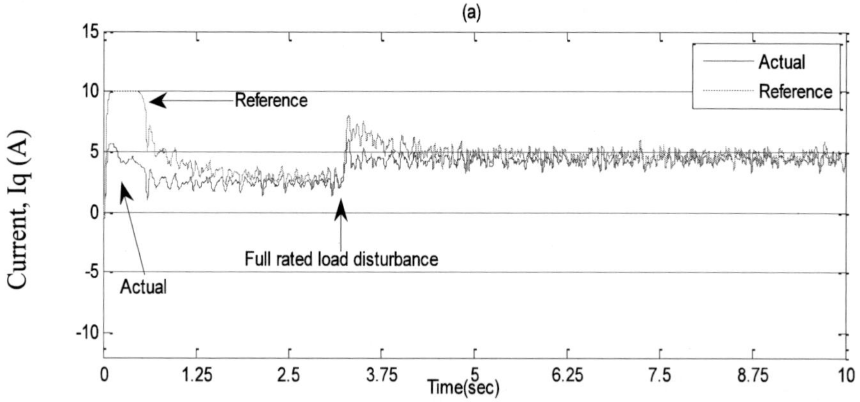

Figure 4.19. (Continued).

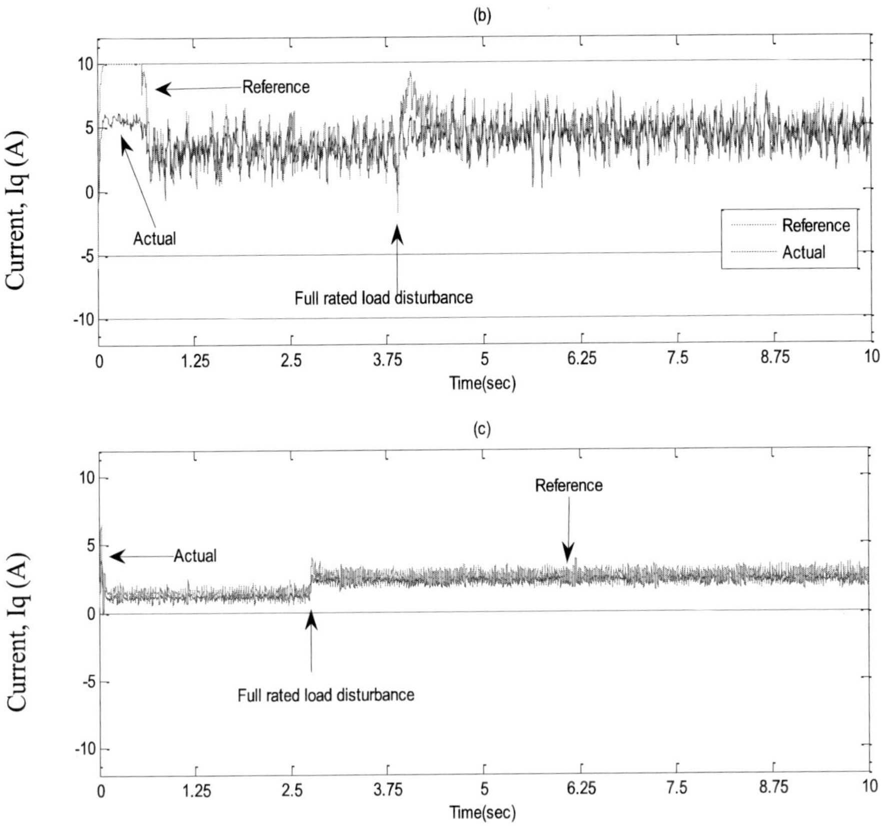

Figure 4.19. Experimental q-axis current responses-based IM drive in Case 2; (a) the tuned PI controller, (b) the conventional BLFC controller, and (c) the proposed NBLFC controller.

4.5. Conclusion

A novel boundary layer fuzzy controller-based IFOC of an IM drive has been presented in this chapter. The structure of the proposed controller is based on e smoothing out the control discontinuity in a thin boundary layer near the sliding surface. The proposed fuzzy system based on the variable boundary layer has been employed to adjust the thickness of boundary layer near the sliding surface to improve tracking performance under small uncertainties. To eliminate the chattering despite large uncertainties, an integral filter has been used in the variable thin boundary layer so that the stability of the proposed NBLFC is guaranteed. The proposed NBLFC-based

IM drive has been successfully implemented in real-time using an ezdspF28335 DSP board and physical sensors for a prototype 1.5 HP motor. The appropriate utilization of the current and position sensors has been contemplated. The performance of the proposed NBLFC has been tested in both simulation and experiment. The performance of the proposed NBLFC controller was found superior to the conventional PI and SMC controllers under different operating conditions such as a step change in command speed, load disturbance and parameter variations over a wide speed range. Furthermore, the proposed NBLFC has reduced the steady-state chattering in current. Thus, the proposed NBLFC ensures less harmonic loss and associated heat dissipation in the motor.

REFERENCES

[1] A. Saghafinia, H. W. Ping, and M. Rahman, "High performance induction motor drive using hybrid fuzzy-pi and pi controllers: A review," *International Review of Electrical Engineering-Iree,* vol. 5, pp. 2000-2012, 2010.

[2] A. Saghafinia and H. W. Ping, "High performance induction motor drive using fuzzy self-tuning hybrid fuzzy controller," in *Power and Energy (PECon), 2010 IEEE International Conference on*, 2010, pp. 468-473.

[3] A. Saghafinia, S. Kahourzade, A. Mahmoudi, W. Hew, and M. N. Uddin, "Broken Rotor Bar Fault Detection of 3-Phase Induction Motor Using Online Adaptive Continuous Wavelet Transform and Fuzzy Logic," *International Review of Electrical Engineering-IREE,* vol. 7, pp. 4383-4394, 2012.

[4] A. Saghafinia, S. Kahourzade, A. Mahmoudi, W. P. Hew, and M. N. Uddin, "On line trained fuzzy logic and adaptive continuous wavelet transform based high precision fault detection of IM with broken rotor bars," in *Industry Applications Society Annual Meeting (IAS), 2012 IEEE*, 2012, pp. 1-8.

[5] M. A. Fnaiech, F. Betin, G. A. Capolino, and F. Fnaiech, "Fuzzy Logic and Sliding-Mode Controls Applied to Six-Phase Induction Machine With Open Phases," *Industrial Electronics, IEEE Transactions on,* vol. 57, pp. 354-364, 2010.

[6] T. Orlowska-Kowalska, M. Dybkowski, and K. Szabat, "Adaptive Sliding-Mode Neuro-Fuzzy Control of the Two-Mass Induction Motor

Drive Without Mechanical Sensors," *Industrial Electronics, IEEE Transactions on,* vol. 57, pp. 553-564, 2010.

[7] L. Viet Quoc, C. Han Ho, and J. Jin-Woo, "Fuzzy Sliding Mode Speed Controller for PM Synchronous Motors With a Load Torque Observer," *Power Electronics, IEEE Transactions on,* vol. 27, pp. 1530-1539, 2012.

[8] Z. H. Salih, K. S. Gaeid, and A. Saghafinia, "Sliding Mode Control of Induction Motor with Vector Control in Field Weakening," *Modern Applied Science,* vol. 9, p. p276, 2015.

[9] A. Saghafinia, H. W. Ping, and M. N. Uddin, "Fuzzy sliding mode control based on boundary layer theory for chattering-free and robust induction motor drive," *The International Journal of Advanced Manufacturing Technology,* vol. 71, pp. 57-68, 2014.

[10] Z. Jinhui, S. Peng, and X. Yuanqing, "Robust Adaptive Sliding-Mode Control for Fuzzy Systems With Mismatched Uncertainties," *Fuzzy Systems, IEEE Transactions on,* vol. 18, pp. 700-711, 2010.

[11] C. Lascu, I. Boldea, and F. Blaabjerg, "A Class of Speed-Sensorless Sliding-Mode Observers for High-Performance Induction Motor Drives," *Industrial Electronics, IEEE Transactions on,* vol. 56, pp. 3394-3403, 2009.

[12] N. B. Cheng, L. W. Guan, L. P. Wang, and J. Han, "Chattering Reduction of Sliding Mode Control by Adopting Nonlinear Saturation Function," *Advanced Materials Research,* vol. 143, pp. 53-61, 2011.

[13] M. L. Tseng and M. S. Chen, "Chattering reduction of sliding mode control by low pass filtering the control signal," *Asian Journal of Control,* vol. 12, pp. 392-398, 2010.

[14] R. Lorenz, "A simplified approach to continuous on-line tuning of field-oriented induction machine drives," *Industry Applications, IEEE Transactions on,* vol. 26, pp. 420-424, 2002.

[15] J. J. E. Slotine and W. Li, *Applied nonlinear control* vol. 461: Prentice hall Englewood Cliffs, NJ, 1991.

[16] F. Cupertino, D. Naso, E. Mininno, and B. Turchiano, "Sliding-Mode Control With Double Boundary Layer for Robust Compensation of Payload Mass and Friction in Linear Motors," *Industry Applications, IEEE Transactions on,* vol. 45, pp. 1688-1696, 2009.

[17] Y. K. Kim and G. J. Jeon, "Error reduction of sliding mode control using sigmoid-type nonlinear interpolation in the boundary layer," *International Journal of Control, and Systems,* vol. 2, pp. 523-529, 2004.

[18] T. Orowska-Kowalska, M. Kaminski, and K. Szabat, "Implementation of a Sliding-Mode Controller With an Integral Function and Fuzzy Gain Value for the Electrical Drive With an Elastic Joint," *Industrial Electronics, IEEE Transactions on,* vol. 57, pp. 1309-1317, 2010.

[19] H. Lee, E. Kim, H. J. Kang, and M. Park, "A new sliding-mode control with fuzzy boundary layer," *Fuzzy Sets and Systems,* vol. 120, pp. 135-143, 2001.

[20] M. Roopaei, M. Zolghadri, and S. Meshksar, "Enhanced adaptive fuzzy sliding mode control for uncertain nonlinear systems," *Communications in Nonlinear Science and Numerical Simulation,* vol. 14, pp. 3670-3681, 2009.

[21] A. Saghafinia, H. W. Ping, M. N. Uddin, and K. S. Gaeid, "Adaptive Fuzzy Sliding-Mode Control Into Chattering-Free IM Drive," *Industry Applications, IEEE Transactions on,* vol. 51, pp. 692-701, 2015.

[22] A. Saghafinia, H. W. Ping, and M. N. Uddin, "Sensored field oriented control of a robust induction motor drive using a novel boundary layer fuzzy controller," *Sensors,* vol. 13, pp. 17025-17056, 2013.

[23] R. Palm, "Sliding mode fuzzy control," 1992, pp. 519-526.

[24] P. Kachroo and M. Tomizuka, "Chattering reduction and error convergence in the sliding-mode control of a class of nonlinear systems," *Automatic Control, IEEE Transactions on,* vol. 41, pp. 1063-1068, 1996.

[25] A. Saghafinia, H. W. Ping, M. N. Uddin, and A. Amindoust, "Teaching of Simulation an Adjustable Speed Drive of Induction Motor Using MATLAB/Simulink in Advanced Electrical Machine Laboratory," *Procedia-Social and Behavioral Sciences,* vol. 103, pp. 912-921, 2013.

[26] P. Franklin and J. D. Powell, "Emami-Naeini. Feedback Control of Dynamic Systems," *Pearson Prentice Hall, New Jersey,* vol. 4, p. 2, 2006.

[27] A. Saghafina, H. W. Ping, M. N. Uddin, and K. S. Gaied, "Adaptive fuzzy sliding-mode control into chattering-free induction motor drive," in *Industry Applications Society Annual Meeting (IAS), 2012 IEEE,* 2012, pp. 1-8.

[28] Q. Cheng and L. Yuan, "Vector Control of an Induction Motor based on a DSP," 2011.

[29] B. Premanode, J. Vongprasert, and C. Toumazou, "Noise Reduction for Nonlinear Nonstationary Time Series Data using Averaging Intrinsic Mode Function," *Algorithms,* vol. 6, pp. 407-429, 2013.

[30] K. H. Eom, S. J. Lee, Y. S. Kyung, C. W. Lee, M. C. Kim, and K. K. Jung, "Improved Kalman Filter Method for Measurement Noise Reduction in Multi Sensor RFID Systems," *Sensors,* vol. 11, pp. 10266-10282, 2011.
[31] T. Y. Poon, N. C. F. Tse, and R. W. H. Lau, "Extending the GMR Current Measurement Range with a Counteracting Magnetic Field," *Sensors,* vol. 13, pp. 8042-8059, 2013.
[32] F. Pan, X. Xiao, Y. Xu, and S. Ren, "An optical AC voltage sensor based on the transverse Pockels effect," *Sensors,* vol. 11, pp. 6593-6602, 2011.

Chapter 5

TEACHING OF SIMULATION AN ADJUSTABLE SPEED DRIVE OF HIGH PERFORMANCE INDUCTION MOTOR USING MATLAB/SIMULINK

ABSTRACT

The simulation of motor complicated applications such as motor speed control conventionally can be a challenge for both undergraduate and postgraduate levels. To easy implementation for several kinds of control structures of an induction motor (IM) drive, some simulators such as MATLAB/Simulink to be necessary-especially for students-to develop and test various motor control algorithms in conducting electrical machines courses. In this chapter, how to teach and simulate an adjustable speed drive of IM using Simulink blocks for an indirect field-oriented control (IFOC) algorithm is presented. The methodology described provides an educational tool in learning improvement of electrical machines course and gives the users the opportunity to immediately apply the information to practical scenarios. The optimized Simulink blocks are used to design the precise Simulink models so that tracking of the speed and flux in the presence of an uncertain load disturbance and parameter variations are provided. The effectiveness of the adjustable IM drive is verified by simulation results at different operating conditions over a wide speed range.

Keywords: MATLAB/Simulink, educational tool, adjustable speed drive, induction motor, indirect field oriented control

5.1. PREAMBLE

To date, academics and practitioners have focused on power electronics education and new computer-aided teaching tools to develop a better learning environment [1, 2]. One of the challenging subjects in these areas is IMs which have been applied extensively due to their advantages such as quiet operation, compact form, simplicity, reliability, low maintenance, economical, rugged, and reliable [3-6]. However, the IM drives are nonlinear, time-varying and complex systems owing to their characteristics [7-10] which require the complex control, circuitry, and inverter over sizing [11]. Finding out of the theory and applications of the IM drives need to some courses including motion control [12], power electronics [13, 14], electric machinery, electric circuits, and electronic courses as well as the related laboratory of these courses. So, the IM drive is developed through some available simulation software to allow students to develop in all of the aforementioned areas in this chapter. The simulation of IM drive, as a teaching tool, supports the classroom teaching as much as its laboratory through the computer-generated graphics, to illustrate easily steady and dynamic state operation of the IM drive under different operation conditions [15]. One of the most widespread simulation software is Mathworks MATLA/Simulink package [16]. The Mathworks MATLAB allows to user for analyzing of complex static systems to be modelled through 'numeric computation and visualization. Also, Simulink allows systems to be simulated dynamically and allows a controller to be modelled with the aid of block diagrams. Moreover, the user is allowed to concentrate on the model, rather than its implementation [17].

This chapter develops a learning environment for IFOC of high performance IM drive using MATLAB/Simulink. Several MATLAB toolboxes, tools, and Simulink Blocksets including power system Blocksets, signal processing toolboxes, fixed-point Blockset, Digital motor control (DMC) library, and IQMath library are used to optimize and design the Simulink model of IM drive. The Simulink model provides tracking of the speed and flux in the presence of an uncertain load disturbance and parameter variations using the real values of the d-axis and q-axis motor currents which utilized to estimate the stator flux angle. The IM drive is fed using voltage source inverter (VSI) which employed the space vector pulse width modulation (SVPWM) technique to optimize the harmonic content of current signal, and switching losses. To show the effectiveness of the simulated adjustable IM drive, simulation results are provided at different operating conditions such as a step change in command speed, load disturbance and parameter variations over a wide speed range [18].

5.2. THE IMPLEMENTED IFOC OF IM DRIVE IN SIMULATION

The discrete time Simulink model with sampling time Ts = $1e^{-4}$ s along with the digital motor control (DMC) and IQMath libraries from TI and Mathworks were used to simulate the IFOC induction motor drive. These libraries were used to optimize the Simulink blocks [18].

The DMC Library provides configurable blocks that can be reused to implement new control strategies. The DMC library is composed of functions represented as blocks. These blocks are categorized as Transforms & Estimators (Clarke, Park, Sliding Mode Observer, Phase Voltage Calculation, and Resolver, Flux, and Speed Calculators and Estimators), Control (Signal Generation, PID, BEMF Commutation, Space Vector Generation), and Peripheral Drivers (PWM abstraction for multiple topologies and techniques, ADC drivers, and motor sensor interfaces). These modules allow users to quickly build, or customize, their own systems. The Library supports the AC motors and comprises both peripheral dependent (software drivers) and target dependent modules.

TI and Mathworks provide a library of highly optimized and high precision math functions in the form of the IQMath library. IQMath is a way of representing a numeric value containing a sign, the integer and the fraction portion in a fixed bit location field. The IQ Math library is available in both fixed-and floating-point versions, enabling easy migration from float to fixed devices. These tools enable developers to quickly determine the processing resources required to implement basic motor control. From this baseline, they are then able to bring in advanced algorithms to trade-off the remaining processing capacity for greater accuracy, better performance, and higher power efficiency, control of multiple motors or integrated power stages like digital power factor correction, and a myriad of other options. In this way, developers can architect systems specifically optimized for their application constraints and requirements.

Parameters of the 1kw 2-pole IM which is used in the simulation are given in Appendix-A. The manufacturing parameters of DC generator are also given in Appendix-B. The SVM-VSI type inverter is modelled based on fast switching IGBTs model from the Simulink toolbox along with the aforementioned libraries in MATLAB. Based on the block diagram of closed-loop vector control of IM drive shown in Figure 3.1, the complete Simulink block diagram is designed which is shown in Appendix-C. Also according to Sections 2.7, 3.3, and 4.3, the Simulink models of the conventional SMC, and the FBLSMC are developed as shown in Appendix-D. The designed Simulink models of the BLFC and the proposed NBLFC speed controllers are also shown in Appendix-E.

5.3. Design of PI-Based Speed and Current Controllers

For comparison purpose, PI controller is designed and implemented in the simulation. The overall block diagram of the basic IFOC of IM drive which uses the PI controller as speed controller is shown in Figure 5.1.

The PI controller is initially tuned by the Ziegler–Nichols method based on stability boundary [19, 20]. It is subsequently tuned through simulations in order to obtain satisfactory responses. The saturation of the controller is avoided by adding a correction of the integral term (*KC*) [19] as shown in Figure 5.2. This method has good load disturbance attenuation. Also, this structure is suggested by TI and MathWorks which are used for PI controllers as speed controller and current controllers in the IFOC of IM drive. The structure of this PI controller is shown in Figure 5.2.

To implement the IFOC of IM drive, the coefficient of PI controllers must be calculated properly. The common approach to calculate these coefficients is trial and error method. However, this method is very time consuming and it is usually impossible to get the proper coefficients for the three PI controllers simultaneously. To overcome this problem according to the suggestion of the TI Company, they are tuned in two stages as follows.

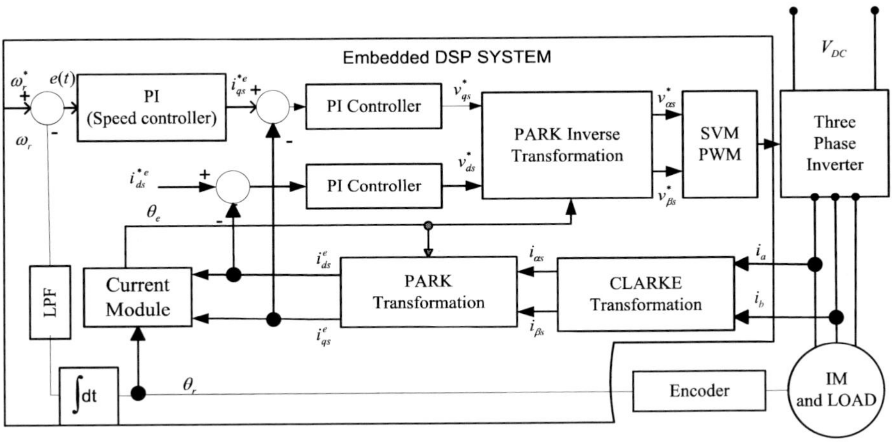

Figure 5.1. The overall block diagram of the basic IFOC of IM drive.

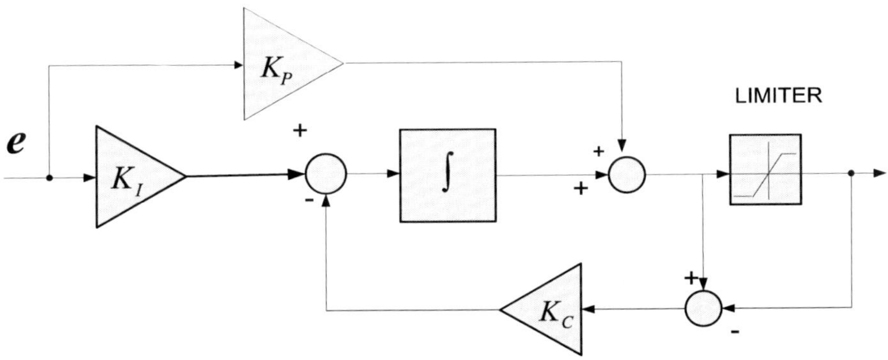

Figure 5.2. PI controller structure with anti-windup correction term.

In the first stage, the coefficients of the two current controllers are tuned without the speed controller (open loop mode) so that the d-axis and q-axis currents follow their commands and the suitable dynamic performances of the d-q-axis currents are obtained. In this stage, the speed is changed by amplitude changes of a ramp controller derived from the DMC library. The d-q-axis currents should be keeping track their commands by amplitude changes of the ramp controller and the stepped commands of the d-axis and q-axis currents. Simulink models, which are changed from closed-loop models for the basic open loop models, are shown in Appendix-F.

In the second stage, the basic IFOC of IM drive is implemented completely and the speed controller gains are only tuned so that the best performances such as minimum overshoot, minimum settling time, and zero steady state error are obtained based on simulation results without any changes in the obtained current controller gains in the previous stage. These gains including the speed and current controllers' gains should be valid for any changes of speed command and load torque. Once the current controller (PI) coefficients are tuned, those remain fixed for any kind of speed controller based IM drive.

5.4. THE SWITCHING TECHNIQUE OF VSI FED IM DRIVE

The SVPWM as the best method for switching technique [21] is considered for VSI fed IM drive in this work as shown in Figure 5.3. In this method, the insulated gate bipolar transistors (IGBTs) must be controlled so that at no time are both switches in the same leg turned on or else the DC supply would be shorted. This requirement may be met by the complementary operation of the switches within a leg. This leads to eight possible switching vectors for the inverter, V0 through V7 with six active switching vectors and two zero vectors as shown in Table 5.1.

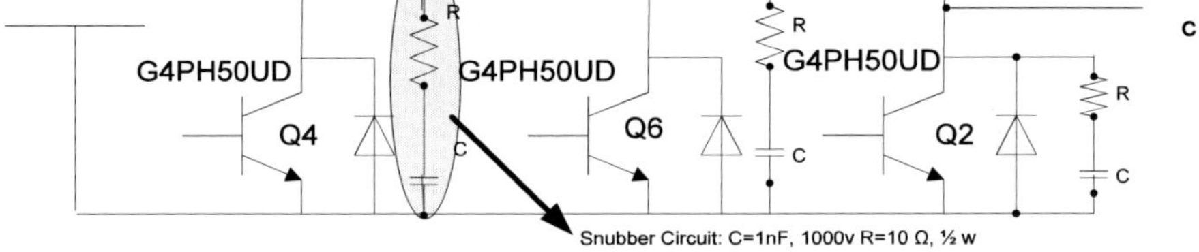

Figure 5.3. VSI fed inverter of the IM drive.

Table 5.1. SVPWM technique switching

Vector	Q1	Q3	Q5	Q4	Q6	Q2	VAB	VBC	VCA
$V_0 = \{000\}$	OFF	OFF	OFF	ON	ON	ON	0	0	0
$V_1 = \{100\}$	ON	OFF	OFF	OFF	ON	ON	+VDC	0	− VDC
$V_2 = \{110\}$	ON	ON	OFF	OFF	OFF	ON	0	+ VDC	− VDC
$V_3 = \{010\}$	OFF	ON	OFF	ON	OFF	ON	− VDC	+ VDC	0
$V_4 = \{011\}$	OFF	ON	ON	ON	OFF	OFF	− VDC	0	+ VDC
$V_5 = \{001\}$	OFF	OFF	ON	ON	ON	OFF	0	− VDC	+ VDC
$V_6 = \{101\}$	ON	OFF	ON	OFF	ON	OFF	+VDC	− VDC	0
$V_7 = \{111\}$	ON	ON	ON	OFF	OFF	OFF	0	0	0

5.5. The IM Drive Overview and its Modelling

The digital motor control (DMC) suggested by TI and Mathworks Companies are used to optimize the Simulink models. The DMC Library provides configurable blocks that can be reused to implement new control strategies. The DMC library is composed of functions represented as blocks. TI and Mathworks Companies also provide a library of highly optimized and high precision math functions in the form of the IQMath library. These libraries allow the engineers/students to quickly build, or customize, their own systems. These libraries along with the discrete time Simulink libraries are employed to generate the IFOC of IM drive [18]. The designed IM drive which configured using discrete-time Simulink with a sampling frequency 10 KHz for the tuned PI controller based IM drive are shown in Figure 5.4 [6]. Parameters of the laboratory 1kw 2-pole IM are given in Appendix-A. To prepare the simulation of the whole system some consideration must be taken into account as follows:

- Since the Simulink models in the MATLAB library are designed carefully, the existing induction motor model in the power system library is used to simulate the induction motor in the proposed model.
- Since the fixed-point blocks (DMC and IQMath libraries) along the float-point blocks are used in the simulated model, the “Data Type Conversion” block must be employed to connect the mentioned blocks.

- To implement the external load and speed reference command, the existing simulation signal in “Signal Processing” library along with manual switches are used to make different shapes of them as shown in Figure 5.4.
- To change the sampling frequency from 10 KHz for each part of the model, the “Rating Transition” block must be employed.
- To make delay in simulation, the “Delay” block (1/z) is used.
- To implement the VSI inverter, the ideal switches are applied so that the switches are excited using the SVPWM technique as described in the previous section.
- The sensors including the rotary optical encoder as speed sensor and transducer as current sensor are modeled using a constant coefficient in accordance with their characteristics.

As shown in Figure 5.4, a subsystem block is applied to generate the three-phase inverter using 600 Vdc which its details are shown in Figure 5.5. The duty cycle of this inverter is controlled using the vector control Simulink model which shown in continue.

”Vector Control” block as the main part of the model (see Figure 5.4) is shown in Figure 5.6. As can be seen in Figure 5.6, the currents in ABC coordinates are converted into the rotating reference frame of d-q axes coordinates through Clarke and park’s transformation from the DMC library. Since the prototype system of the model employees an encoder 5000puse/revolution, the rotor position obtained from the motor is scaled in pulse resolution by multiplying 4*5000/2Π. Based on the measured rotor position, the rotor speed is computed using numerical backward differentiation with sampling time 1 kHz which its detail is shown in Figure 5.7. Then according to the actual and command speed, the q-axis reference current is calculated. Considering the nominal value for the d-axis reference current, the current controllers calculate the d-axis and q-axis reference voltages as can be seen in Figure 5.8. It is worthy to note that the “saturation” subsystem (see Figure 5.9) is considered to prevent the motor to be in saturation mode. The detail of this subsystem is shown in Figure 5.9(a).

The six SVM pulses for the inverter switches are made through the inverse park’s transformations, space vector generator, and “PWM Scaling” block as can be seen in Figure 5.9(b). The “PWM Scaling” block is employed to adjust the switching frequency of SVM-PWM for 18 KHZ by multiplying its input by 12 using “IQN1 to IQN2” block as shown in Figure 5.9(b).

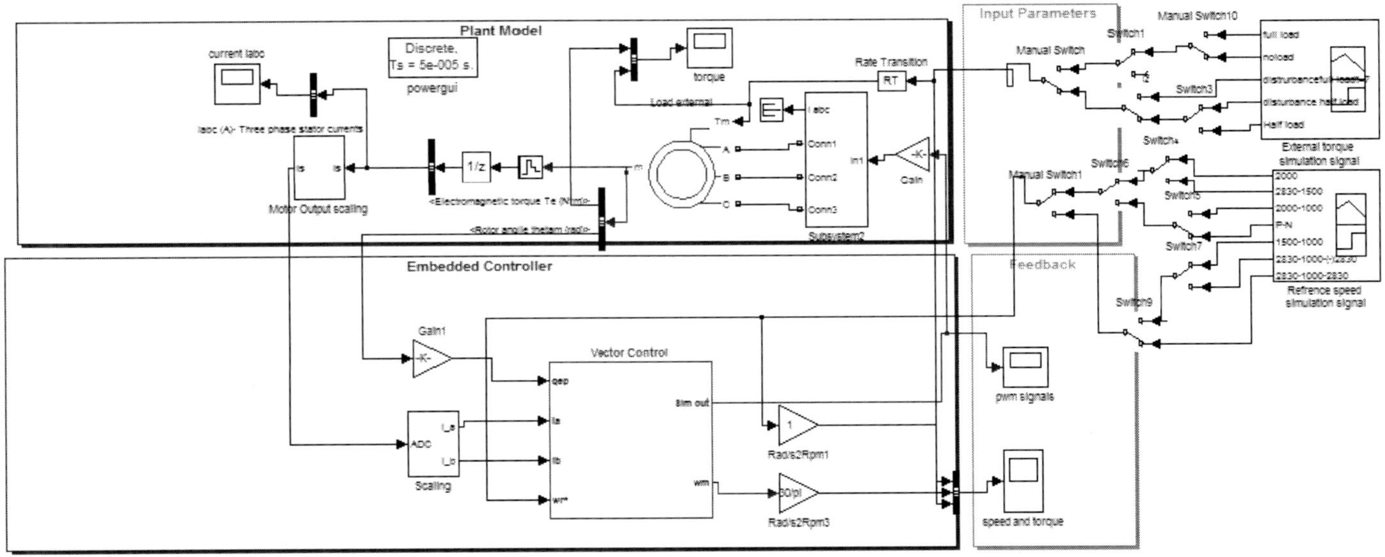

Figure 5.4. The main Simulink model for the IFOC of IM drive.

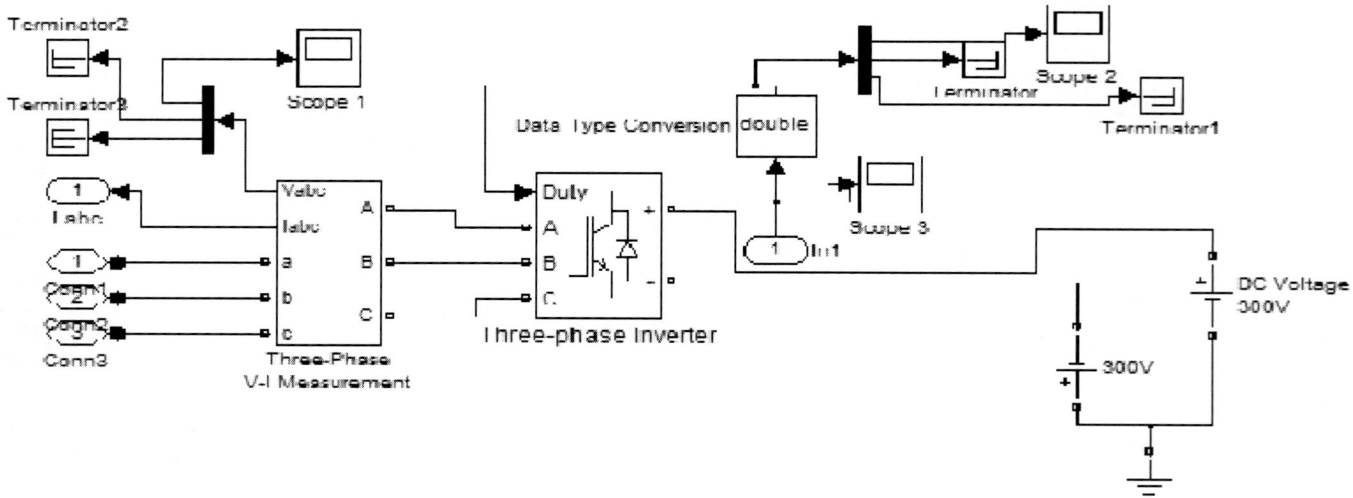

Figure 5.5. The Simulink Model of Ideal inverter.

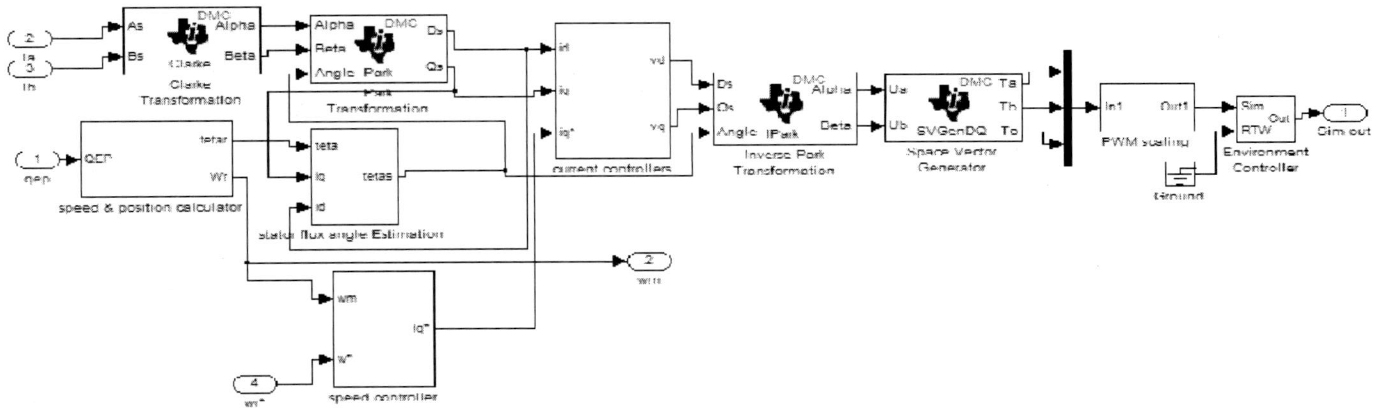

Figure 5.6. The Simulink Model of Vector control to generate the space vectors.

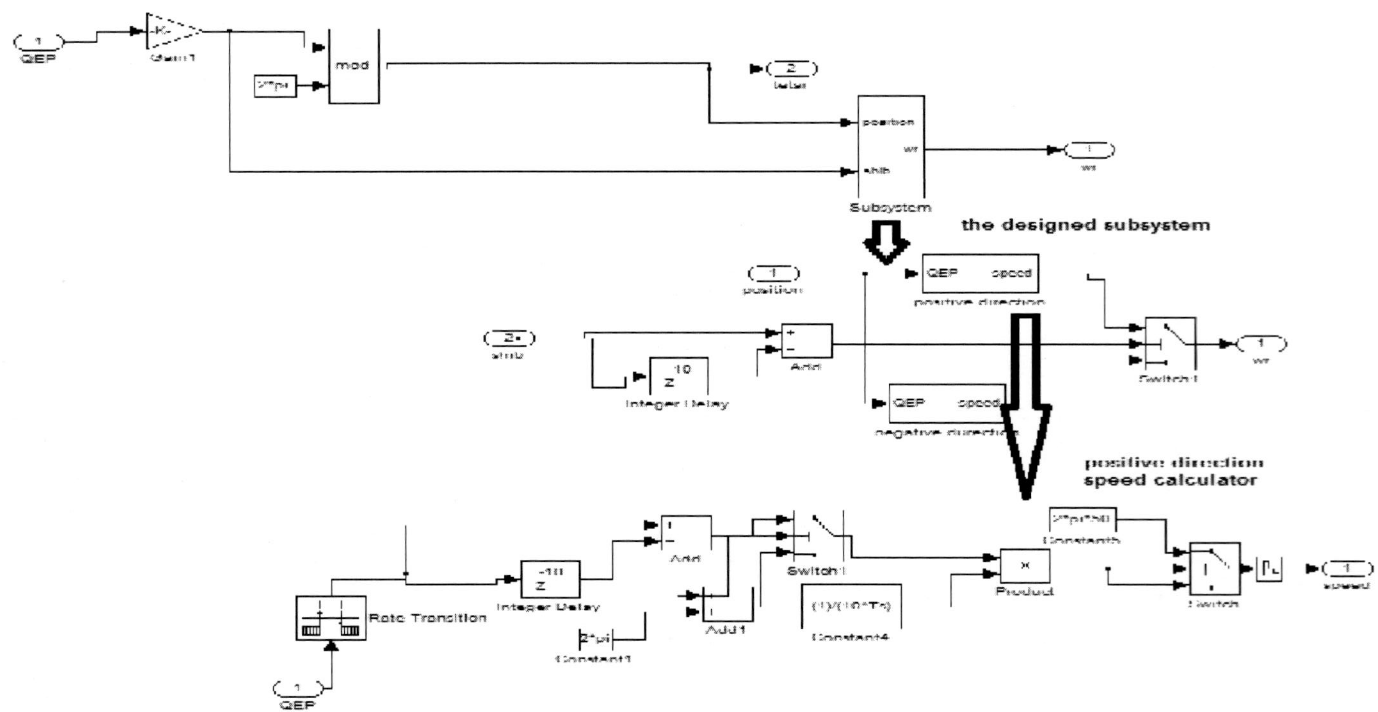

Figure 5.7. The Simulink model of speed calculation.

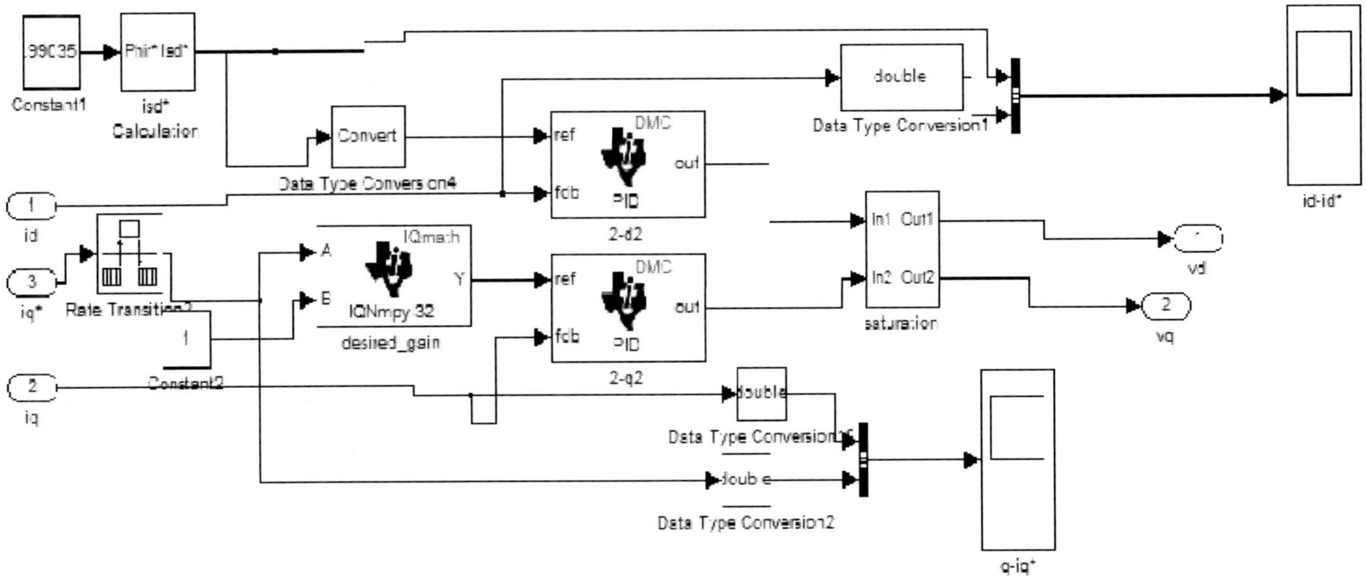

Figure 5.8. The Simulink Model of the PI current controllers.

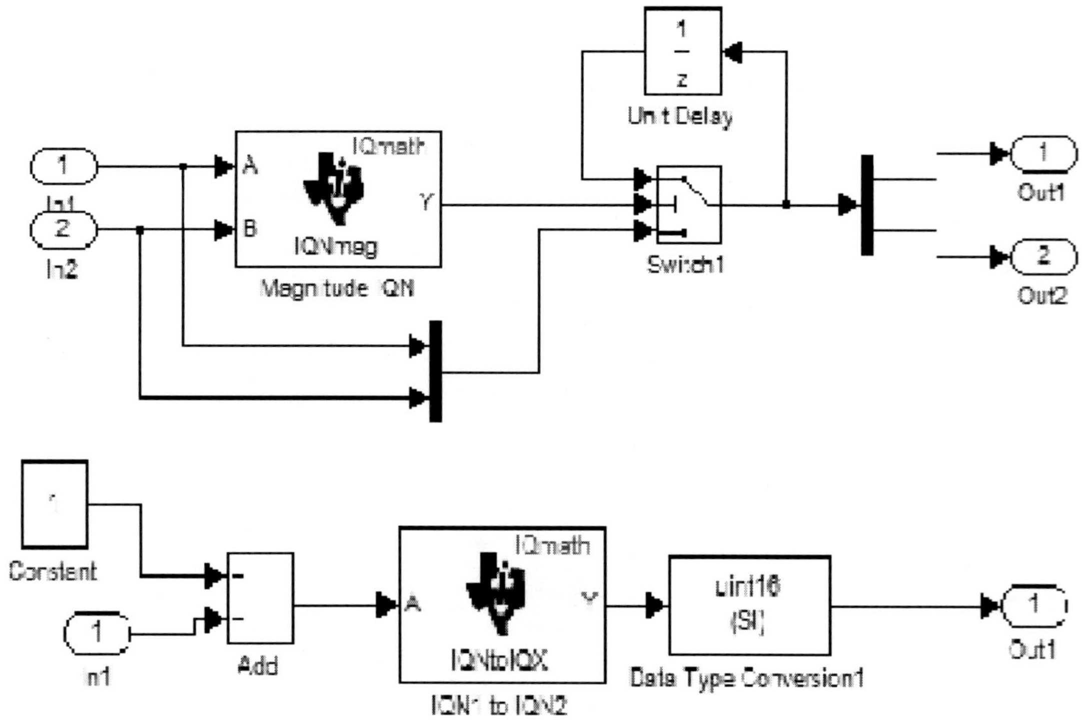

(a)

(b)

Figure 5.9. The Simulink Model. (a) “Saturation block,” (b) “PWM scaling” block.

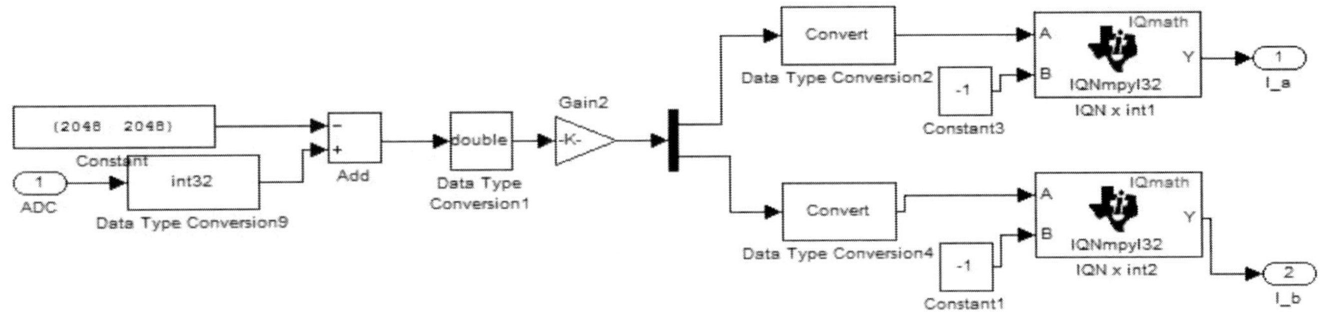

Figure 5.10. The Simulink model of current measurement.

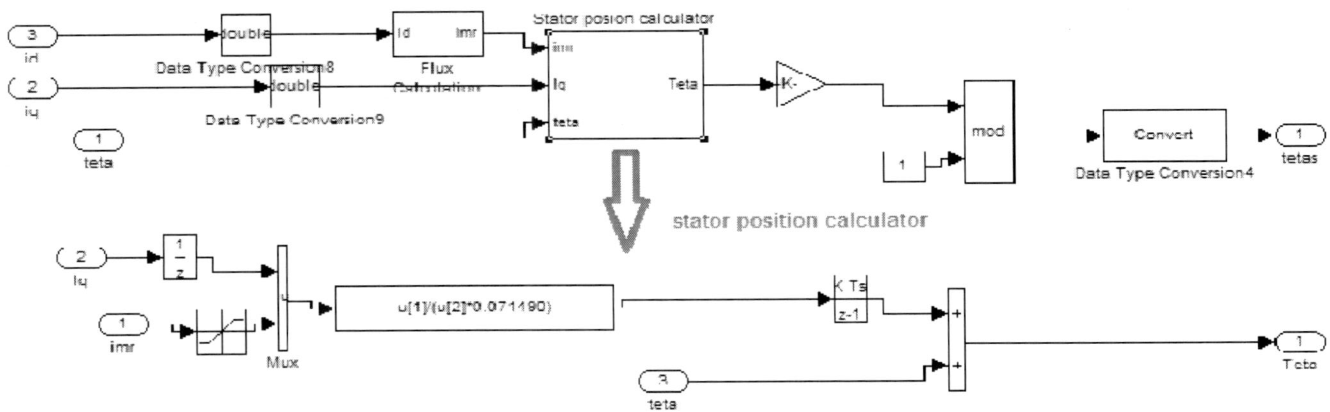

Figure 5.11. The Simulink Model of stator flux angle estimator.

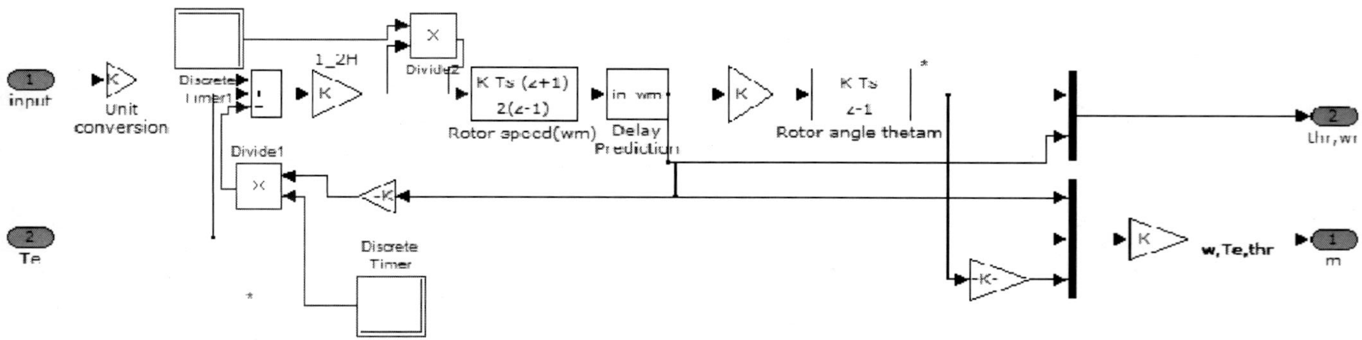

Figure 5.12. The changed mechanical model of the motor modeled.

The motor currents (I_a, I_b) calculated from the motor is scaled (by multiplying their inputs by 2048/10 and the getting results plus by 2048) to match the transducer sensor inputs which applied in the prototype system. It is worth to note that the maximum current measured by transducers is 10A. These currents are given to the "scaling" block (see Figure 5.4) to prepare the suitable value (by multiplying their inputs by 10/2048 and the getting results subtract from 2048) in ABC coordinates as its details can be seen in Figure 5.10.

To implementation IFOC algorithm properly, the real values of the d-axis and q-axis motor currents are applied to estimate the stator flux angle as can be seen in Figure 5.11. The accuracy of this algorithm depends on the accuracy of some motor parameters such as rotor time constant which obtained for the model equal to 0.07149 as shown in Figure 5.11.

After implementation of the IFOC algorithm, the laboratory 1Kw motor parameters are put in the model by clicking on the motor in the main model (Figure 5.4). All parts of the motor model can be seen by right clicking on the motor and choosing the "look under musk." The modelled motor parameters also can be changed by choosing "link options" and "break link" by right clicking on the part of the motor which must be changed. The mechanical model is changed according to the aforementioned instructions to change the inertia and friction factors as shown in Figure 5.12, for example.

5.6. Simulation Results

For simulation tests, following cases including parameter variations and external load disturbance are considered. If not mentioned, other parameters are considered nominal in all the cases for the tuned PI. The PI parameters are found as, Kp = 0.3, Ki = 0.0001, and Kc = 0.0001.

Case 1: Step changes in command speed with no load and full load.

Case 2: Step increase in load from '0' to 50% and '0' to 75% of rated load at t = 7s.

Case 3: Stator resistance is chosen two times of nominal value applied at t = 7sec while the motor starts with half rated load.

Case 4: Inertia and friction coefficient are increased three times of nominal value at t = 7s while half rated load is applied from the beginning.

For case 1, simulation results are illustrated in Figure 5.13(a)-(d) and 5.14(a)-(d). From Figure 5.13(a) and 5.14(a), it can be seen that the tuned PI shows the acceptable speed response in the case of the dynamic and steady-state performance. It can also be seen from Figure 5.13(c, d) and 5.14(c, d) that the suitable tracking response of speed and flux has been obtained for the tuned PI which verifies the suitable performance of the implemented IFOC algorithm.

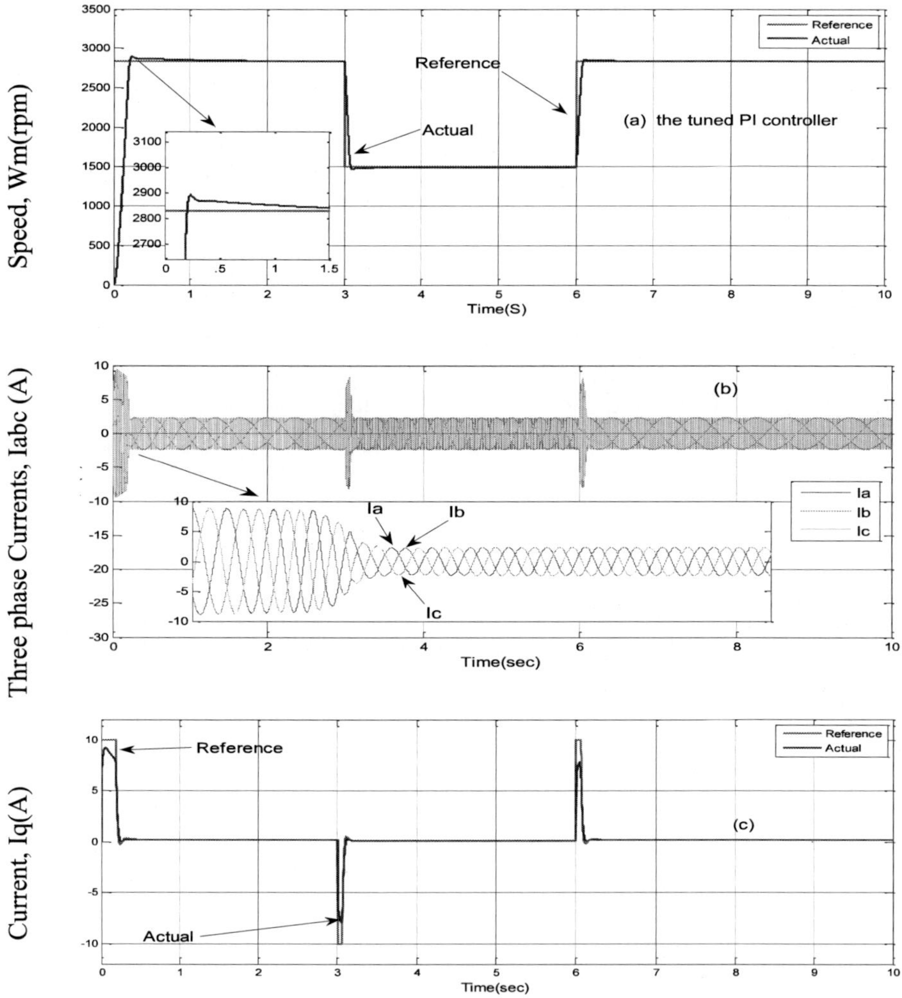

Figure 5.13. (Continued).

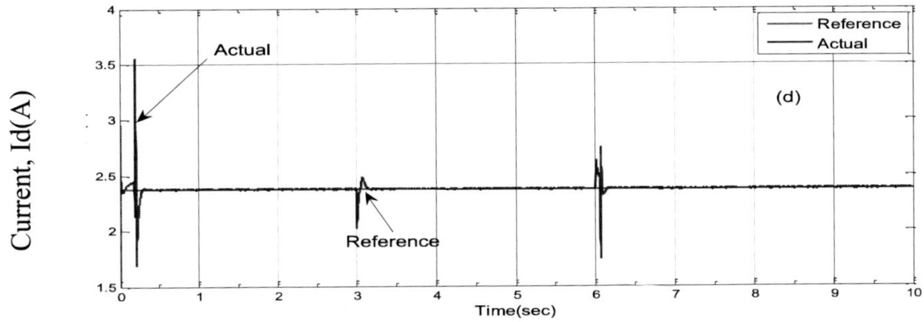

Figure 5.13. Simulated response of the tuned PI controller based IM drive at no load in Case 1; (a) speed (b), 3-phase currents (Iabc), (c) q-axis current, and (d) d-axis current.

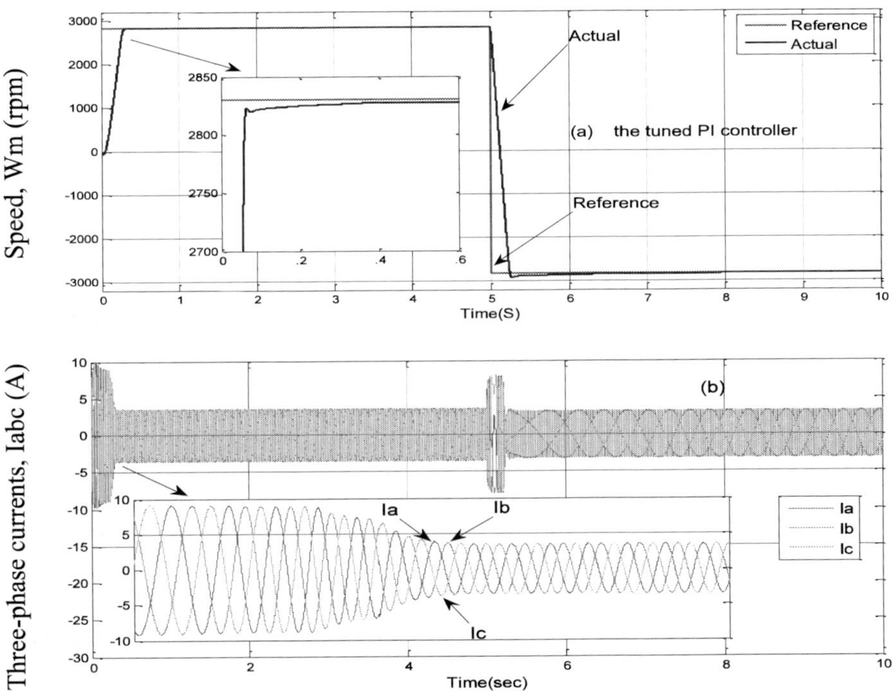

Figure 5.14. (Continued).

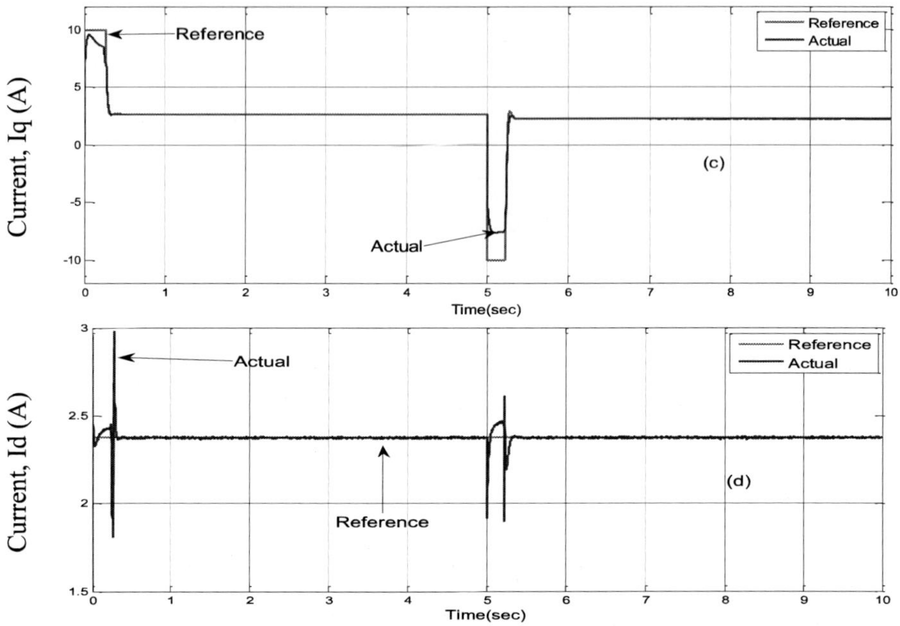

Figure 5.14. Simulated response of the tuned PI controller based IM drive at full rated load in Case 1; (a) speed, (b) 3-phase currents (Iabc), (c) q-axis current, and (d) d-axis current.

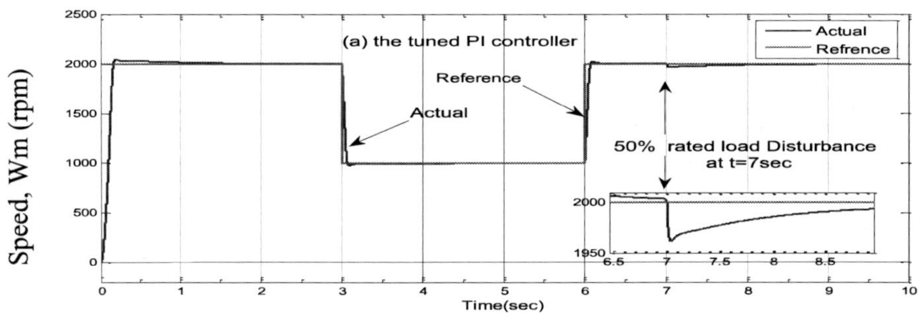

Figure 5.15. (Continued).

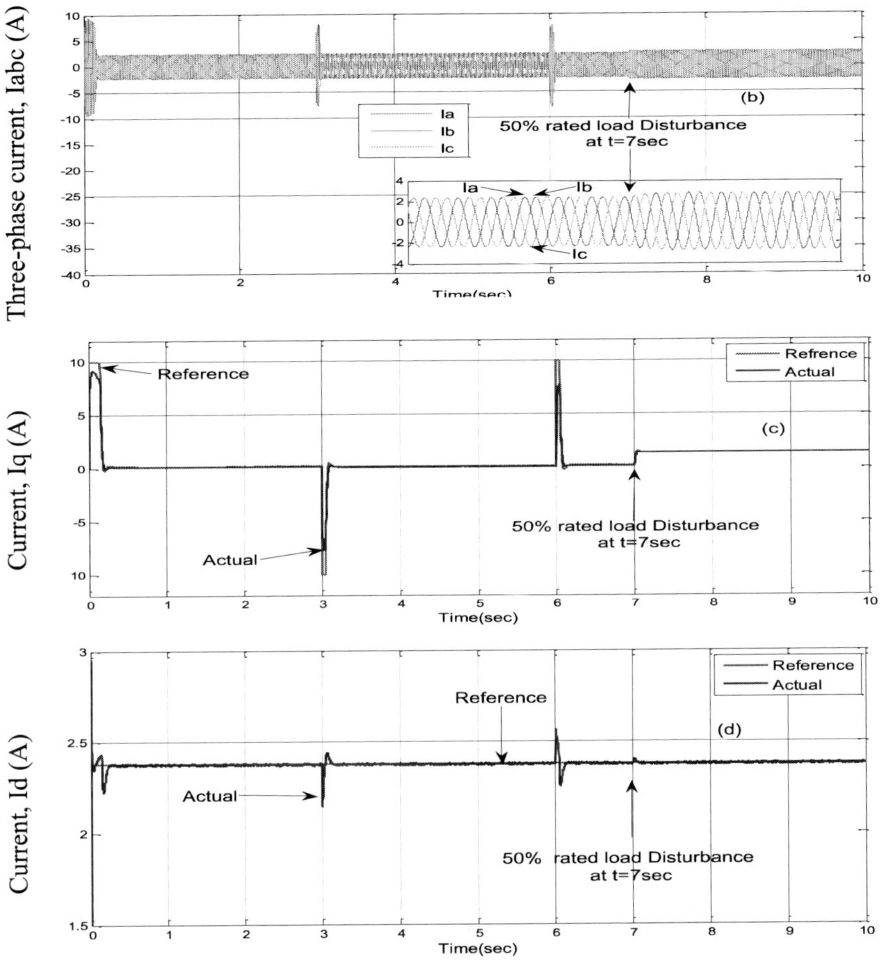

Figure 5.15. Simulated response of the tuned PI controller based IM drive with 50% rated load disturbance at t = 7sec in Case 2; (a) speed, (b) 3-phase currents (Iabc), (c) q-axis current, and (d) d-axis current.

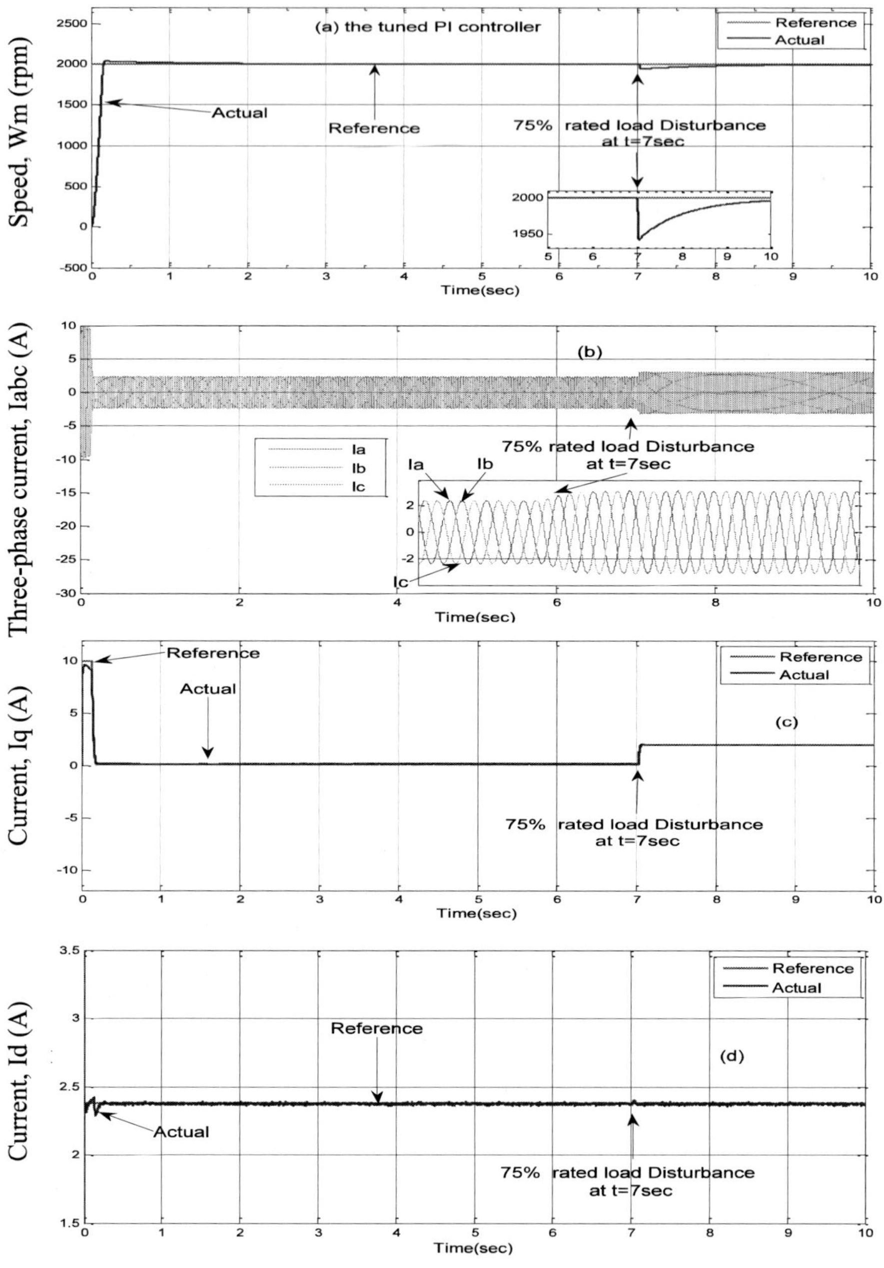

Figure 5.16. Simulated response of the tuned PI controller based IM drive with 75% rated load disturbance at t = 7sec in Case 2; (a) speed, (b) 3-phase currents (Iabc), (c) q-axis current, and (d) d-axis current.

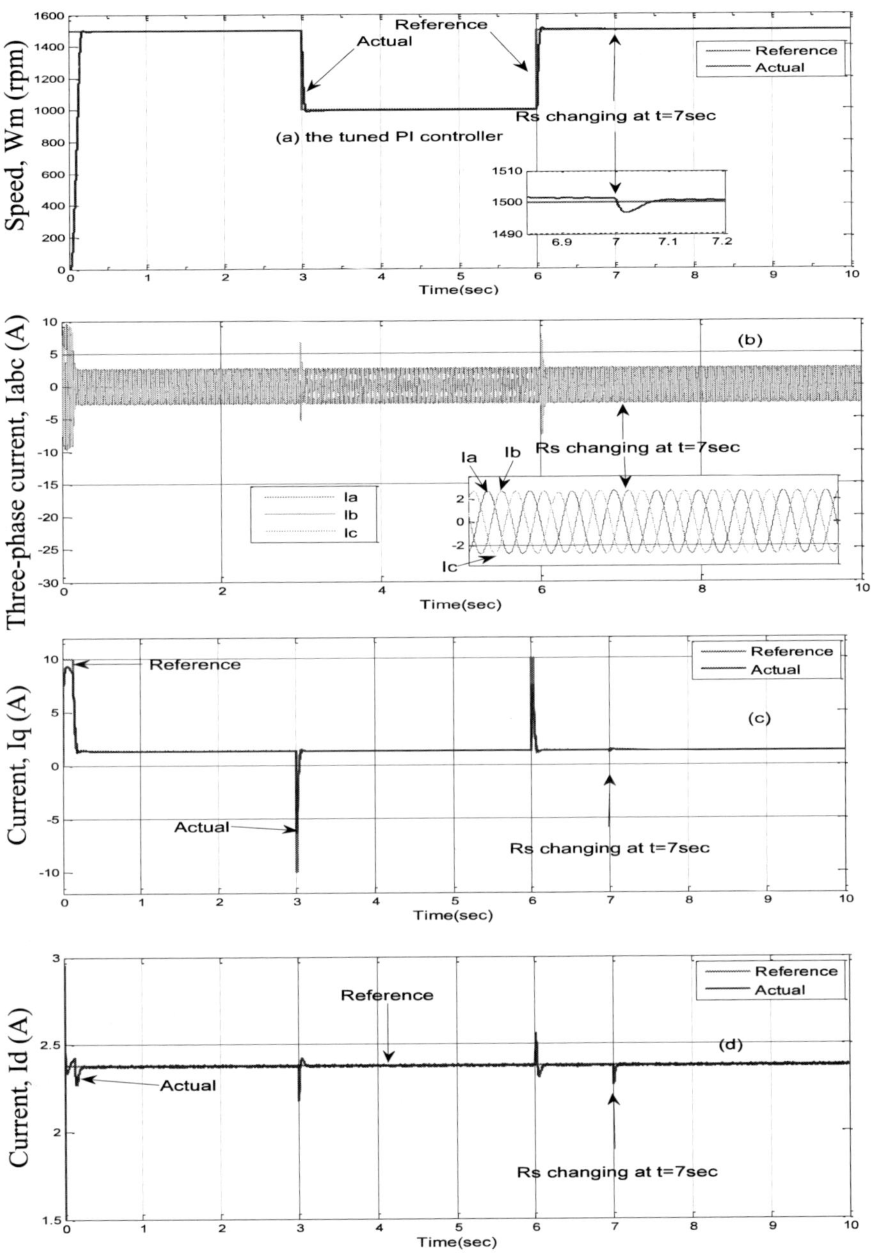

Figure 5.17. Simulated response of the tuned PI controller based IM drive with Rs variation at t = 7sec in Case 3. (a) speed, (b) 3-phase currents (Iabc), (c) q-axis current, and (d) d-axis current.

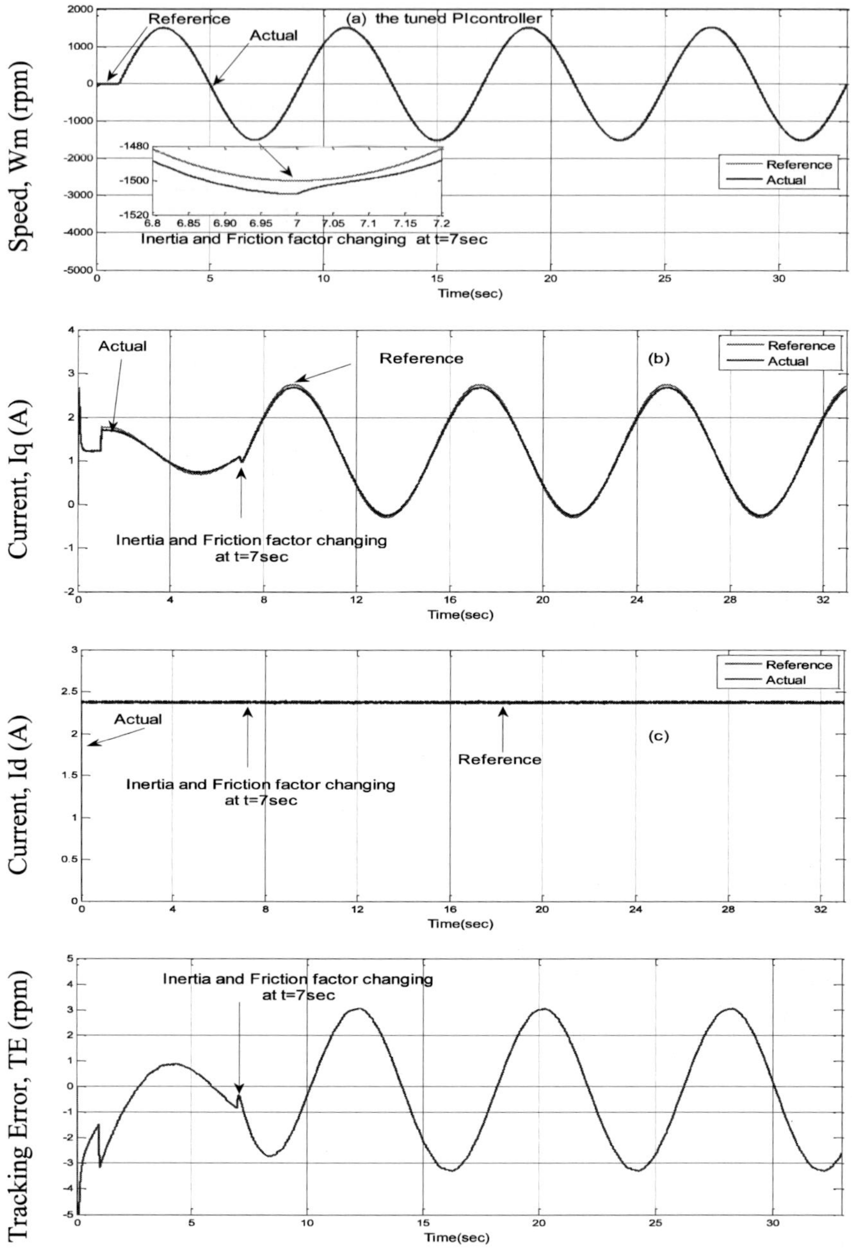

Figure 5.18. Simulated response of the tuned PI controller based IM drive with Jr and Br variations at t = 7 in Case 4; (a) speed, (b) q-axis current, (c) d-axis current, and (d) Tracking error.

For Case 2, simulation results are shown in Figure 5.15(a)-(d) and 5.16(a)-(d). From Figure 5.15(a) and 5.16(a), it can be seen that the PI controller shows a significant dip in speed($\approx$ 40 rpm for 50% and 60 rpm for 75% rated load) when the step increase in load is applied at t = 7s. It is also found that the tuned PI follows the q-d-axis commands despite the large uncertainties which shows the suitable performance of the IFOC algorithm in this case.

In Case 3, Simulation results are shown in Figure 5.17(a)-(d). It is found from Figure 5.17 that the PI shows the suitable responses in spite of the parameter variations and it has acceptable tracking performance in both speed & current.

In Case 4, the sinusoidal command is selected as speed command to show the tracking error properly. The other parameter variations are also tested in this case. Simulation results are shown in Figure 5.18(a)-(d) for the tuned PI. It is found from Figure 5.18 that the PI shows the suitable responses in spite of the parameter variations and it has acceptable tracking performance in both speed & current. Thus, the high accuracy of the implemented IFOC algorithm is obtained despite uncertainties such as external load disturbance and parameter variations in the simulated model.

5.7. CONCLUSION

A simulation model of an adjustable speed drive of IM using the optimized Simulink blocks for an IFOC algorithm has been presented in this chapter. The work presented has verified to be effective and valid for measuring IM drive parameters. Also, an educational tool in learning improvement of electrical machines course through the Simulink models for adjustable speed drive of 1 kW laboratory IM has provided in the method described. Moreover, all aspects of simulation implementation have been explored through the Simulink to give the student the opportunity to easy implement the IM drive. The system has been found to be well suited for simulation learning of the various motor control algorithms for both undergraduate and postgraduate levels. The simulation results verify the effectiveness of the model.

REFERENCES

[1] N. Mohan, W. P. Robbins, P. Imbertson, T. M. Undeland, R. C. Panaitescu, A. K. Jain, et al., "Restructuring of first courses in power electronics and electric drives that integrates digital control," *Power Electronics, IEEE Transactions on,* vol. 18, pp. 429-437, 2003.

[2] A. Keyhani, M. N. Marwali, L. E. Higuera, G. Athalye, and G. Baumgartner, "An integrated virtual learning system for the development of motor drive systems," *Power Systems, IEEE Transactions on,* vol. 17, pp. 1-6, 2002.

[3] A. Saghafinia, H. W. Ping, and M. A. Rahman, "High Performance Induction Motor Drive Using Hybrid Fuzzy-PI and PI Controllers: a Review," *International Review of Electrical Engineering-Iree,* vol. 5, pp. 2000-2012, Sep-Oct 2010.

[4] A. Saghafinia, H. Ping, and M. Uddin, "Designing Self-Tuning Mechanism On Hybrid Fuzzy Controller For High Performance And Robust Induction Motor Drive," *the International Journal of Advanced Technology & Engineering Research,* vol. 3, 2013.

[5] A. Saghafinia and H. W. Ping, "High performance induction motor drive using fuzzy self-tuning hybrid fuzzy controller," in *Power and Energy (PECon), 2010 IEEE International Conference on*, 2010, pp. 468-473.

[6] A. Saghafinia, H. W. Ping, M. N. Uddin, and K. S. Gaeid, "Adaptive Fuzzy Sliding-Mode Control Into Chattering-Free IM Drive," *Industry Applications, IEEE Transactions on,* vol. 51, pp. 692-701, 2015.

[7] L. Qaseer, S. Purushothaman, and F. de Leon, "Closed-Form Analysis of Squirrel-Cage Induction Motors With Anisotropic Modeling of Stator and Rotor," *Energy Conversion, IEEE Transactions on,* vol. PP, pp. 1-8, 2012.

[8] A. Saghafinia, S. Kahourzade, A. Mahmoudi, W. Hew, and M. N. Uddin, "Broken Rotor Bar Fault Detection of 3-Phase Induction Motor Using Online Adaptive Continuous Wavelet Transform and Fuzzy Logic," *International Review of Electrical Engineering-Iree,* vol. 7, pp. 4383-4394, 2012.

[9] A. Saghafinia, S. Kahourzade, A. Mahmoudi, W. P. Hew, and M. N. Uddin, "On line trained fuzzy logic and adaptive continuous wavelet transform based high precision fault detection of IM with broken rotor

bars, in *Industry Applications Society Annual Meeting (IAS), 2012 IEEE*, 2012, pp. 1-8.

[10] A. Saghafina, H. W. Ping, M. N. Uddin, and K. S. Gaied, "Adaptive fuzzy sliding-mode control into chattering-free induction motor drive," in *Industry Applications Society Annual Meeting (IAS), 2012 IEEE*, 2012, pp. 1-8.

[11] U. Bakshi and V. Bakshi, *Electrical Circuits and Machines*: Technical Publications, 2009.

[12] R. D. Lorenz, T. A. Lipo, and D. W. Novotny, "Motion control with induction motors," *Proceedings of the IEEE,* vol. 82, pp. 1215-1240, 1994.

[13] T. J. Goulart and D. Consonni, "Automated system for measuring electrical three-phase power components," *Education, IEEE Transactions on,* vol. 44, pp. 336-341, 2001.

[14] J. M. Jimenez-Martinez, F. Soto, E. Jodar, J. A. Villarejo, and J. Roca-Dorda, "A new approach for teaching power electronics converter experiments," *Education, IEEE Transactions on,* vol. 48, pp. 513-519, 2005.

[15] L.-H. Hoang, "Modeling and simulation of electrical drives using MATLAB/Simulink and Power System Blockset," in *Industrial Electronics Society, 2001. IECON '01. The 27th Annual Conference of the IEEE*, 2001, pp. 1603-1611 vol.3.

[16] S. Ayasun and C. O. Nwankpa, "Induction motor tests using MATLAB/Simulink and their integration into undergraduate electric machinery courses, " *Education, IEEE Transactions on,* vol. 48, pp. 37-46, 2005.

[17] J. Rodriguez-Resendiz, G. Herrera-Ruiz, and E. A. Rivas-Araiza, "Adjustable speed drive project for teaching a servo systems course laboratory," *Education, IEEE Transactions on,* vol. 54, pp. 657-666, 2011.

[18] A. Saghafinia, H. W. Ping, M. N. Uddin, and A. Amindoust, "Teaching of Simulation an Adjustable Speed Drive of Induction Motor Using MATLAB/Simulink in Advanced Electrical Machine Laboratory," *Procedia-Social and Behavioral Sciences,* vol. 103, pp. 912-921, 2013.

[19] P. Franklin and J. D. Powell, "Emami-Naeini. Feedback Control of Dynamic Systems," *Pearson Prentice Hall, New Jersey,* vol. 4, p. 2, 2006.

[20] B. C. Kuo, *Automatic control systems*: Prentice Hall PTR Upper Saddle River, NJ, USA, 1981.

[21] B. K. Bose, *Modern power electronics and AC drives*: Prentice Hall PTR USA, 2002.

[22] B. Akin; and M. Bhardwaj, "Sensored Field Oriented Control of 3-Phase Induction Motors," 2010.

Chapter 6

Real-Time Implementaion of High Performance Induction Motor Drive Using Digital Signal Processor Board TI TMS320F28335

Abstract

Digital Signal Processors (DSPs) and their programming are main parts in motor control applications, which can be extremely challenging for students or any first-time users. The Digital Signal Processors (DSPs) help users to easy implementation for several kinds of control structures of induction motor (IM) drives for the real-time Simulink model as it does not need to write the programming due to the advantage of the real-time workshop (RTW). This chapter presents the detail real-time implementaion of high performance induction motor drive using digital signal processor board TI TMS320F28335. MATLAB/Simulink environment, controls and analyzes the system and is runed from Spectrum Digital through a PC on an ezdspF28335 platform.

Keywords: digital signal processor, MATLAB/Simulink, real-time workshop, Induction motor drive, hardware implementaion, software implementaion, practicing engineers

6.1. PREAMBLE

The electrical motors convert more than 50% of electrical energy into mechanical energy. Over the years, induction motors (IM's) have been widely used in industry for variable speed drives due to some of the advantages such as robust construction, low cost, low maintenance, and reliability [1-3]. However, the IM drives are nonlinear, time-varying and complex systems due to their characteristics [4-6], which require the complex control, circuitry, and inverter over sizing [7]. Therefore, the intelligent controller demands special attention for high performance IM drives [8-12]. In this chapter, the intelligent IM drive is developed through the real-time Simulink to allow students /first-time users to implement a rapid prototyping of indirect field oriented control of high performance IM drive.

To reduce the design cycle of new DSP application, fast development tools such as a real time workshop (RTW) of MATLAB/Simulink is necessary. Particularly, RTW is very useful for students to develop and test various real-time control algorithms or DSP algorithms in conducting DSP courses [13-15]. This work offers a rapid prototyping system using RTW of MATLAB/Simulink and the Texas Instruments (TI) DSP TMS320F28335. After an extensive simulation investigation in chapters 3, 4 on the proposed novel fuzzy theory and sliding-mode speed control based IM drives and the teaching it to apply high perormance IM drive in simulation in chapter 5, the hardware and software implementation of IFOC based IM drive is described step by step to give the first-time user/student an opportunity to apply this information to specific scenarios rapidly. The methodology described in this chapter can also be useful in developing similar hardware or software for applications in industrial and test equipment. Moreover, some practical issues and software implementation of the control algorithm are presented in this chapter.

6.2. EXPERIMENTAL SETUP

The experimental setup for the real-time implementation of the proposed speed controller is shown in Figure 6.1 and Figure 6.2. In these Figures, some labels are chosen to show details of the experimental setup as follows. To implement load for IM, a DC generator (G) has been coupled with the tested IM which is labeled as "M" so that some resistors (L) as DC generator's load has been connected to the output voltage terminals of the DC generator. The DC generator has been also excited by a DC voltage controller, which labeled as "DVC." To measure the speed of the motor, an optical quadrature incremental encoder (QEP) has been used so that is directly connected to the rotor shaft to measure the rotor position. A speed and torque controller (STC) has been also used for testing and measuring speed and torque of IM to check the hardware setup performance. Current transducers as current sensors have been used to measure the motor actual currents. An interface circuit (IC) has been designed to match the rotor position and motor currents signals with the QEP and A/D and channels of DSP board, respectively.

A gate drive circuit (GD) is designed to increase the power level of the firing pulses for the insulated gate bipolar transistor (IGBT) switches of the designed VSI drive inverter (INV) and it also isolates between low power control and high power supply circuit. In fact, the designed gate drive is matched with the PWM output channels of DSP. The details of the designed inverter along with snubber circuits and the gate drive are shown in Appendix-G.

Figure 6.1. Experimental setup of the proposed IM drive system.

The power circuits include 3-phase variable AC autotransformer (AT), power supply (PS), rectifier (R), and an IGBT inverter (INV). A large capacitor (2×2200μf) (C) is used to smooth of the output DC voltage of the rectifier. Two DC power supplies (PDC1, PDC2) are used for making +5V, -5V, +15V, -15V voltages. The ezdspF28335 board is labeled as "DSP" as shown in Figure 6.2. A digital storage oscilloscope (O) is used to capture the analog current signal of the ADC channels of ezdspF28335 board. Current meters and voltage meters are used to read current and voltage of the IM and DC generator, respectively. The drive, which has been established through both hardware and software, is described below.

6.3. Hardware Implementaion for IM Drive

The block diagram of the hardware schematic of VSI fed IM drive is shown in

Figure 6.3. As shown in

Figure 6.3, the ezdspF28335 board is used to implement the IM drive. The ezdspF28335 is a stand-alone card, and is used to evaluate the TMS320F28335 digital signal controller (DSC) and determine if it meets the application requirements. Also, it is an excellent platform to develop and run software for the TMS320F28335 processor.

As shown in

Figure 6.3, the ezdspF28335 includes the following hardware and software features: (1) TMS320F28335 digital signal controller; (2) 150 MHz operating speed; (3) on chip 32-bit floating point unit; (4) 68K bytes on-chip RAM; (5) 512K bytes on-chip Flash memory; (6) 256K bytes off-chip SRAM memory; (7) On chip 12 bit Analog to Digital (A/D) converter with 16 input channels; (8) 30 MHz input clock; (9) On board RS-232 connector with line driver; (10) On board CAN 2.0 interface with line driver and connector; (11) Multiple Expansion Connectors (analog, I/O); (12) On board embedded USB JTAG Controller; (13) 5-volt only operation with supplied AC adapter; (14) On board IEEE 1149.1JTAG emulation connector; (15) TI F28xx Code Composer Studio integrated development environment (IDE) Version 3.3; (16) Texas Instruments' flash APIs to support the F28335; (17) Texas Instruments' F28335 header files and example software. In addition, in this figure, the TMS320F28335 DSP as a 32-bit C2000 family of the TI's DSP is suitable to execute complex control algorithms along with the right mix of peripherals to interface with the various components of the DMC hardware like the ADC, ePWM, QEP, eCAP, etc. According to the characteristics of C2000 family DSP, it, along with the advanced control scheme, provides the system improvements as follows.

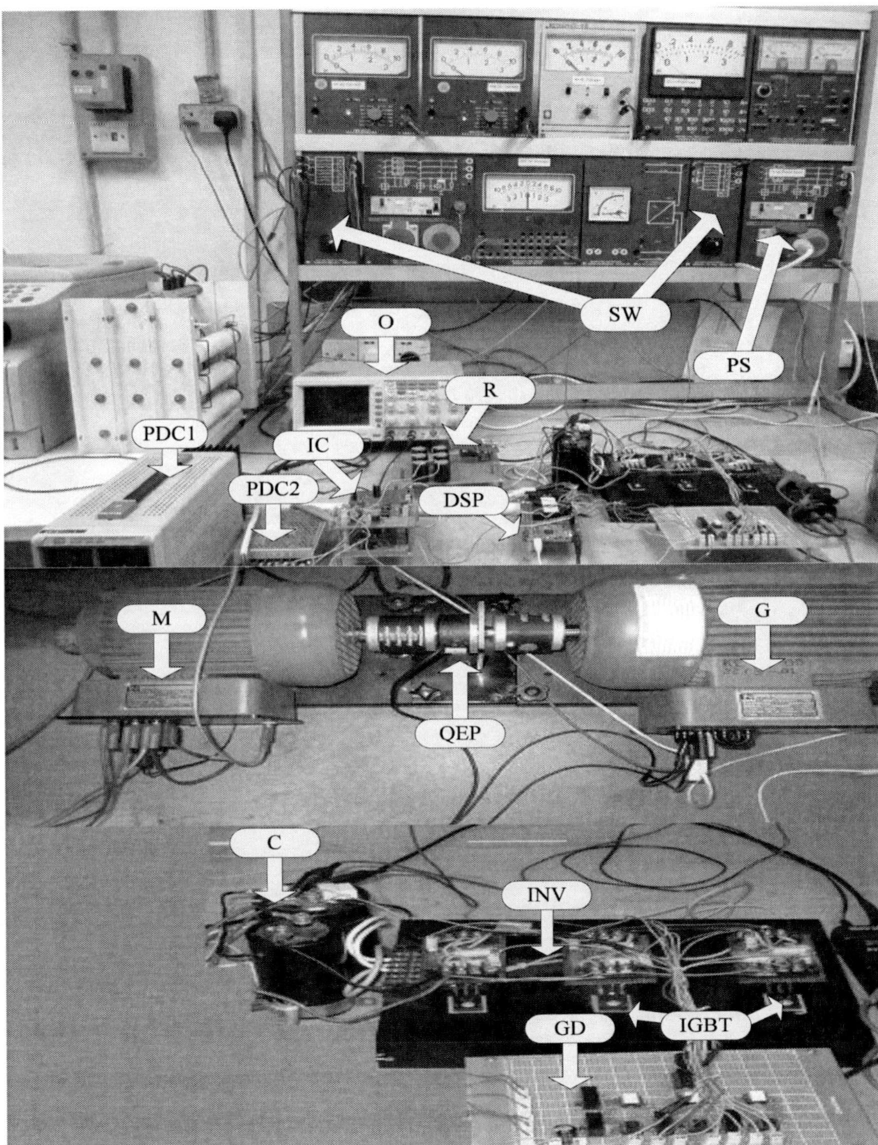

Figure 6.2. Experimental setups of the proposed speed controller of the IM drive (zoom-in view).

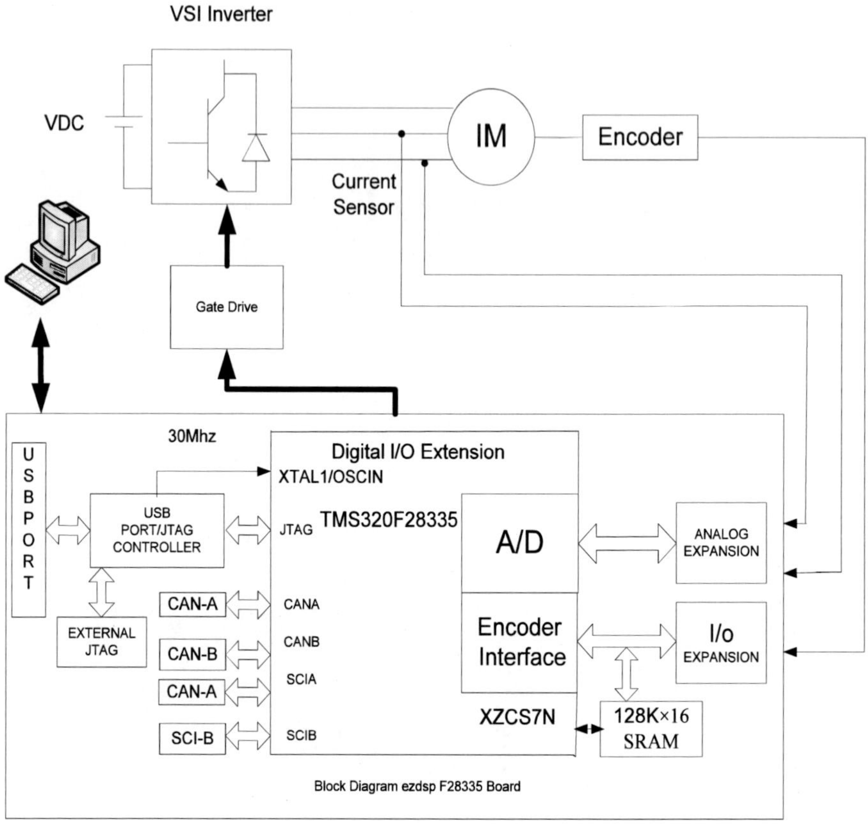

Figure 6.3. Block diagram of the hardware schematic for real-time implementation of VSI fed IM drive.

- Favors system cost reduction by an efficient control in all speed range implying proper dimensioning of power device circuits.
- Using advanced control algorithms it is possible to reduce torque ripple, thus resulting in lower vibration and longer lifetime of the motor
- Advanced control algorithms reduce harmonics generated by the inverter thus reducing filter cost.
- Use of sensorless algorithms eliminates the need for speed or position sensor.
- Decreases the number of look-up tables, thus, which reduces the amount of memory required.

- The real-time generation of smooth near-optimal reference profiles and motion trajectories, results in better performance.
- Generation of high resolution PWM's is possible with the use of ePWM peripheral for controlling the power switching inverters.
- Provides single chip control system.
- Enables control of multi-variable and complex systems using modern intelligent methods such as neural networks and fuzzy logic.
- Performs adaptive control C2000 controllers with speed capabilities to concurrently monitor the system and control it. A dynamic control algorithm adapts itself in real time to variations in system behavior.
- Performs parameter identification for sensorless control algorithms, self-commissioning, online parameter estimation update.
- Performs advanced torque ripple and acoustic noise reduction.
- Provides diagnostic monitoring with spectrum analysis. By observing the frequency spectrum of mechanical vibrations, failure modes can be predicted in the early stages.
- Produces sharp-cut-off notch filters that eliminate narrow-band mechanical resonance. Notch filters remove energy that would otherwise excite resonant modes and possibly make the system unstable.

As shown in Figure 6.3, rotor position is sensed by an optical incremental encoder which is mounted on the rotor shaft and is provided as feedback to ezdspF28335 through the I/O expansion. The used encoder generates 5000 pulses per revolution which is multiplied by 4 by using the quadrature mode. Therefore, the resolution of the encoder is 0.018°. A 16-bit position counter is used to count the encoder data and it is reset in each revolution. The motor speed is calculated from the measured rotor position angles using a discrete difference equation. According to Figure 6.3, the actual motor currents are measured by the transducer sensors, which have a current range of ±10A. The current signals are fed back to ezdspF28335 board through A/D channels. The output current signal of these sensors is converted to a voltage across the resistor connected between the output terminal of the sensor and ground. The interface circuit (IC) is used to match the rotor position and motor currents signals with the QEP and A/D and channels of ezdspF28335 board, respectively and it also reduces the noises. The generated six SVM-PWM pulses from the digital I/O extension of ezdspF28335 board are used as firing pulses for the inverter switches through the gate drive circuit.

6.4. Softeare Implementaion for IM Drive

6.4.1. Preamble

Quite often two computers are used in a host/ target configuration to implement real-time control systems. The host is a computer without real-time requirements, which develops environment, data visualization and control panel in the form of a Graphic User Interface (GUI) residing within it. The target as a second computer or an embedded system based on a DSP board is established by the real-time system. A block diagram of real time control and simulation including real-time tasks, on-line task, and off-line task is shown in Figure 6.4. The main idea behind using the real time control is to make a smooth transition from the non-real(or offline) analysis and simulation of the real time experiments and implementation [16, 17]. Control strategies, structures and parameter values are required for the real time simulation. To perform the simulations and real-time experiments in a structured and simple manner, the integration of real-time control and simulation environment are the best method for the designer.

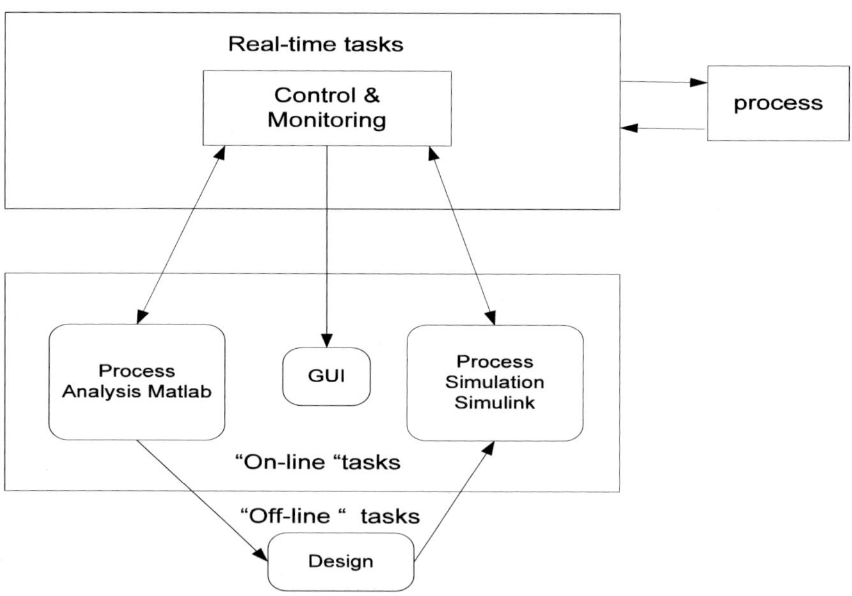

Figure 6.4. Block diagram for real-time control system.

In the case of electrical machine drives, the basic real time control system can be presented in Figure 6.5. As shown in Figure 6.5, the connection indicate a closed loop between various entities involved in receiving signals from sensors and ultimately translating those to the inverter gating circuit pulses through the controller.

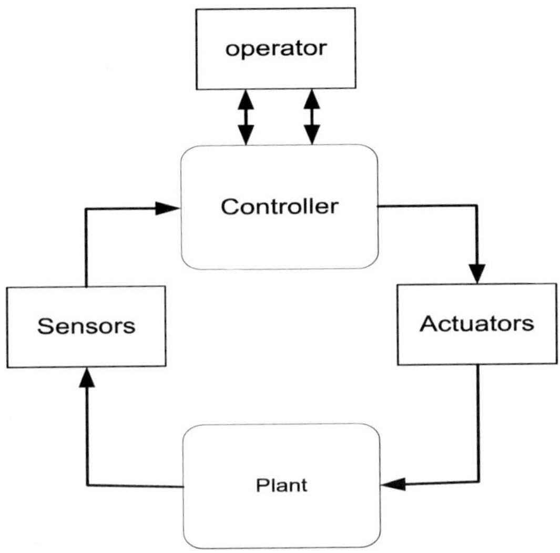

Figure 6.5. Block diagram of real time control for electrical drives.

6.4.2. The Sotware Structure

According to real-time control process for IM drive, a real-time Simulink model of the complete drive is developed to implement the control algorithm as shown in Appendix-H. The real time system structure provides the implementation of advanced motor drive control algorithms to evaluate their performance in real time. Thereafter, the model is downloaded to the ezdspF28335 board through MATLAB and Code Composer Studio (CCStudio) software. A PC as GUI interface is used to control and monitor the data in an online manner. The sampling frequency used in this work for the proposed NBLFC and the proposed AFSMC controller based IM drive is 4 KHz. The switching frequency used is also 9 KHz for two proposed controller based IM drive to prevent the 'overrun error', which indicates too much computational burden for the processor. Figure 6.6 shows the flow chart of software for real-time implementation, which is provided using download

MATLAB Simulink in DSP through CCStudio software automatically. After initializing all the required variables in the DSP blocks of TMS320F28335 in MATLAB library, the time interrupt routine is set up to read the current values and rotor position angle every 250μs for the two proposed controllers based IM drive. It is noted that initializing some the required variables is done in the target preference of the DSP blocks, which must be put in the main page of the real-time Simulink model. The motor currents calculated through ADC block from the DSP blocks in MATLAB Simulink channels are converted appropriately to obtain the actual current values in real-time Simulink. After that, these digitalized currents in ABC coordinates are converted into the rotating reference frame of d-q axes coordinates through Clarke and Park's transformation from the DMC library. On the other hand, the rotor position angle is measured by encoder, which can be calculated in radian by the 2Π/(5000× 4). Based on the measured position angle, the rotor speed is computed using numerical backward differentiation with sampling time 1 kHz. Thereafter, according to the calculated speed and the designed proposed speed controller mentioned in chapter 3, 4, the q-axis reference current is calculated. The d-axis reference current is chosen to be 60% of nominal d-axis motor current considering the nominal value of DC link. By choosing the d-axis reference current, the current controllers PID from the DMC library calculate the d-axis and q-axis reference voltage by using initial parameters stored in memory. The six SVM pulses for the inverter switches are constructed through DMC blocks such as the inverse Park's transformations and space vector generator. To prevent the shorting of the dc bus voltage to ground, all off-to-on transitions of the SVM pulses are delayed by the dead time of 0.2 milliseconds and they are sent to inverter gate drive through the EPWM blocks of the DSP blocks and digital I/O extensions of the ezdspF28335 board.

6.4.3. The Data Saving on DSP's Memory

Since the ezdspF28335 board does not have any digital to analog (D/A) converter, it is unable to acquire and display output variables such as, d-axis and q-axis current, motor speed, and etc. On the other hand, the program is made automatically by downloading MATLAB compiled execute files into the DSP and it is hard to change some part of the downloaded software. So to save output variables in the DSP's memory (Zone 7 of DSP) and reading it by CCStudio software in on-line manner, some extra Simulink models have been applied in real-time Simulink for the first time. It is done using the "Memory

Copy" block of the DMC blocks as shown in Appendix-G. It is worthy to note that 4000 data samples have been considered for each parameter to save them in Zone 7 of DSP. Since the program has been executed for 10 second, the sampling frequency has been selected at 400 Hz to save the data.

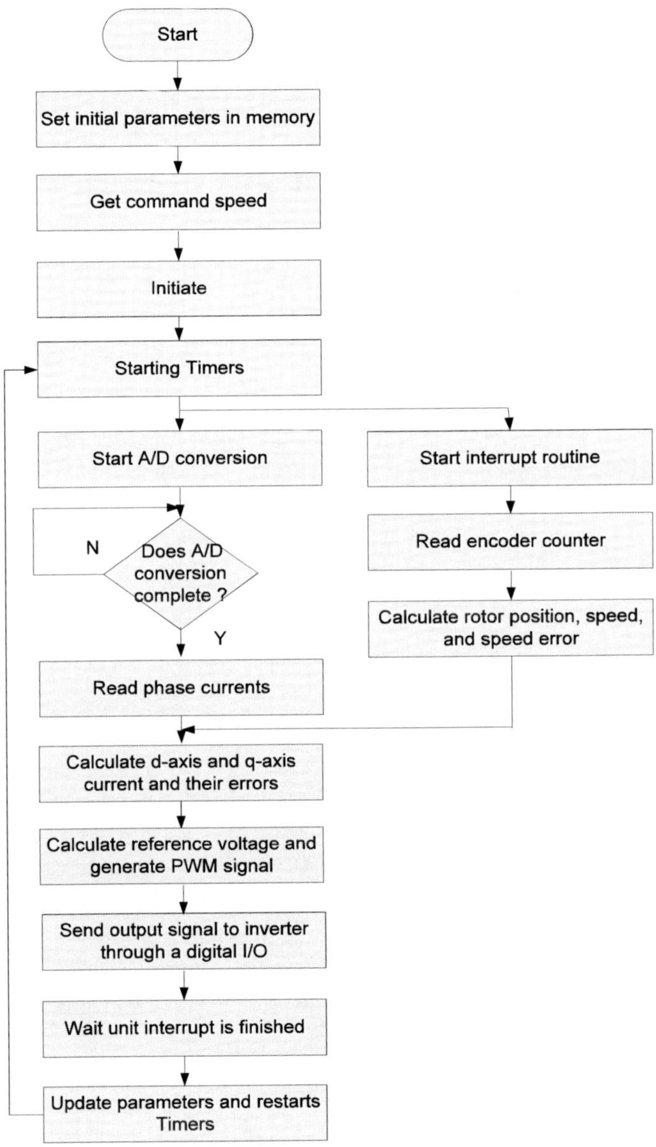

Figure 6.6. The flow chart of the software for real-time implementation of IM drive.

6.5. CONCLUSION

The detail DSP based real-time implementation procedure of the proposed intelligent speed controllers for IM drive has been presented in this chapter. All aspects of the hardware and software implementation have been explored through a real-time Simulink model for an ezdspF28335 platform. Thus, this chapter has made the first-time users/students/practicing engineers an opportunity for easier and faster implementation of the intelligent controller based vector control of IM drive. This chapter could be used in learning DSP based laboratory implementation of adjustable speed drives through the real-time Simulink model.

REFERENCES

[1] A. Saghafinia, H. W. Ping, and M. A. Rahman, "High Performance Induction Motor Drive Using Hybrid Fuzzy-PI and PI Controllers: a Review," *International Review of Electrical Engineering-Iree,* vol. 5, pp. 2000-2012, Sep-Oct 2010.

[2] Z. H. Salih, K. S. Gaeid, and A. Saghafinia, "Sliding Mode Control of Induction Motor with Vector Control in Field Weakening," *Modern Applied Science,* vol. 9, p. p276, 2015.

[3] A. Saghafinia and H. W. Ping, "High performance induction motor drive using fuzzy self-tuning hybrid fuzzy controller," in *Power and Energy (PECon), 2010 IEEE International Conference on*, 2010, pp. 468-473.

[4] L. Qaseer, S. Purushothaman, and F. de Leon, "Closed-Form Analysis of Squirrel-Cage Induction Motors With Anisotropic Modeling of Stator and Rotor," *Energy Conversion, IEEE Transactions on,* vol. PP, pp. 1-8, 2012.

[5] A. Saghafinia, S. Kahourzade, A. Mahmoudi, W. P. Hew, and M. N. Uddin, "On line trained fuzzy logic and adaptive continuous wavelet transform based high precision fault detection of IM with broken rotor bars," in *Industry Applications Society Annual Meeting (IAS), 2012 IEEE*, 2012, pp. 1-8.

[6] A. Saghafinia, S. Kahourzade, A. Mahmoudi, W. Hew, and M. N. Uddin, "Broken Rotor Bar Fault Detection of 3-Phase Induction Motor Using Online Adaptive Continuous Wavelet Transform and Fuzzy

Logic," *International Review of Electrical Engineering-iree,* vol. 7, pp. 4383-4394, 2012.

[7] U. Bakshi and V. Bakshi, *Electrical Circuits and Machines*: Technical Publications, 2009.

[8] A. Saghafinia, H. W. Ping, and M. N. Uddin, "Fuzzy sliding mode control based on boundary layer theory for chattering-free and robust induction motor drive," *The International Journal of Advanced Manufacturing Technology,* vol. 71, pp. 57-68, 2014.

[9] A. Saghafinia, H. Ping, and M. Uddin, "Designing Self-Tuning Mechanism on Hybrid Fuzzy Controller for High Performance and Robust Induction Motor Drive," *the International Journal of Advanced Technology & Engineering Research,* vol. 3, 2013.

[10] A. Abbondanti, *Method of flux control in induction motors driven by variable frequency, variable voltage supplies,* 1977, pp. 177–184.

[11] A. Amindoust, S. Ahmed, and A. Saghafinia, "Supplier Selection and Quota Allocation Decisions Under Uncertainty: Review and Future Research Directions."

[12] A. Saghafina, H. W. Ping, M. N. Uddin, and K. S. Gaied, "Adaptive fuzzy sliding-mode control into chattering-free induction motor drive," in *Industry Applications Society Annual Meeting (IAS), 2012 IEEE,* 2012, pp. 1-8.

[13] W. S. Gan, Y. K. Chong, W. Gong, and W. T. Tan, "Rapid prototyping system for teaching real-time digital signal processing," *Education, IEEE Transactions on,* vol. 43, pp. 19-24, 2000.

[14] A. Saghafinia, H. W. Ping, M. N. Uddin, and K. S. Gaeid, "Adaptive Fuzzy Sliding-Mode Control Into Chattering-Free IM Drive," *Industry Applications, IEEE Transactions on,* vol. 51, pp. 692-701, 2015.

[15] A. Saghafinia, H. W. Ping, and M. N. Uddin, "Sensored field oriented control of a robust induction motor drive using a novel boundary layer fuzzy controller," *Sensors,* vol. 13, pp. 17025-17056, 2013.

[16] W. Grega, K. Kolek, and A. Turnau, *Rapid prototyping environment for real-time control education,* 1998, pp. 85-92.

[17] P. Menghal and A. J. Laxmi, *Real time control of electrical machine drives: A review*, 2010, pp. 1-6.

Chapter 7

CONCLUSION AND SUGGESTION FOR FUTURE WORK

ABSTRACT

This chapter presents the summary of the contribution and the concluding remarks of this work. In addition, the originality of the work is explored in the next section. Furthermore, some suggestions for the future works and studies are explored in this chapter. Finally, all related Appendices are listed at the end of this book.

Keywords: conclusion, future works, originality of the work, work suggestions

7.1. SUMMARY OF THE WORK

Since high performance electrical motor drives are essential requirements for industrial drive applications, the development of such motor drives become important engineering challenge to the researchers. As discussed in chapter 1, the indirect field oriented control is very widely used to achieve high performance motor drives as many advanced algorithms and strategies can be applied by this method. However, changes in parameters and disturbances such as external load torque and unmodeled dynamics often degrade the electrical drive performance. Therefore, in order to achieve high performance IFOC based IM drive sophisticated speed control method is required.

As presented in chapter one, unlike the linear methods, variable-structure control (VSC) or in particular, sliding–mode control (SMC) is applied in IM drives due to some important characteristics such as robustness to system parameter and fast dynamic response. To reduce the influence of the mathematical model and omit the chattering problem, the SMC based controllers are applied along with the intelligent controllers.

In chapter two, a literature review of IFOC of IM drive has been presented in detail. From this review, it is observed that the use of VSI inverters along with the PWM techniques with regard to parameters such as cost, efficiency, and transient state are the best option in high performance IM drives applications. Due to high computing capabilities provided by the advanced DSPs, they are applied to implement the high performance of the SVM-VSI fed IM drives. A literature review of IM drive incorporating different control techniques such as, intelligent and nonlinear speed controllers has also been discussed. From this review, considering that the IM drive often face the possibility of large uncertainties in real-time, it is found that the existing BLFC controller is not able to eliminate the chattering completely while it improves the tracking performance of the systems with the possibility of large uncertainty. In addition, the available fuzzy sliding-mode controller techniques are not sufficient to eliminate the chattering while the tracking performance is improved. Therefore, in this book, two intelligent speed controllers are developed based on the SMC and fuzzy logic theories to achieve high performance drive. Finally, the theoretical bases that can be used in order to complete a book project are explained in the last section.

As presented in Chapter 3, the proposed NFSMC control for chattering-free and robust IM drive has been explored. For the proposed NFSMC, fuzzy inference system has been assigned for reaching controller part of fuzzy sliding mode controller (FSMC) to eliminate the chattering phenomenon in spite of the small and large uncertainties in the system so that stability of the system is guaranteed. Also, the equivalent control part has been estimated to avoid the computational burden by an averaging filter. On the other hand, the averaging filter assisted to improve of tracking performance despite the possibility of large uncertainties in the system so that the stability of the system is guaranteed [1]. Extensive computer simulations of the conventional SMC, FSMC, and the proposed NFSMC speed controllers have been investigated using MATLAB Simulink and their results have been shown at different operating conditions. From these results, it is found that the the proposed NFSMC speed controller based IM drives are robust to parameter variations and external load disturbance and the chattering is absent in the speed and currents of the proposed controller.

As presented in Chapter 4, the proposed NBLFC control based on the boundary layer approach for an indirect field-oriented control (IFOC) of an induction motor (IM) drive has been explored. For the proposed NBLFC, a fuzzy system has been used to adjust the boundary layer thickness to improve the tracking performance and eliminate the chattering problem under small uncertainties, which known as a BLFC. To eliminate the chattering under the possibility of large uncertainties in real-time, the integral filter has been also proposed inside the variable boundary layer [2]. The real-time implementation of the conventional PI, the BLFC, and the proposed NBLFC based speed controller of IM drives using DSP controller board ezdspF28335 for a laboratory 1 kW motor have been presented. Extensive computer simulations and numerous experimental results for the conventional PI, BLFC, and the proposed NBLFC controllers have been shown in this chapter. A real-time performance comparison of the proposed NBLFC with the conventional PI and BLFC controller has been also presented in this chapter. It is found from experimental and simulation result that the proposed NBLFC based IM drive is robustness to parameter variations and external load disturbance and the chattering is absent in speed and currents. It is also found that the tracking capability, and steady state and dynamic performance of the proposed NBLFC are better than the conventional PI and BLFC controllers.

In the fifth Chapter, how to teach and simulate an adjustable speed drive of IM using Simulink blocks for an indirect field-oriented control (IFOC) algorithm has been presented. The methodology described has been provided an educational tool in learning improvement of electrical machines course and has been given the users the opportunity to immediately apply the information to practical scenarios. The optimized Simulink blocks have been used to design the precise Simulink models so that tracking of the speed and flux in the presence of an uncertain load disturbance and parameter variations has been provided. The effectiveness of the adjustable IM drive has been verified by simulation the tuned PI controller results at different operating conditions over a wide speed range.

In the sixth chapter, the detail real-time implementaion of high performance induction motor drive using digital signal processor board TI TMS320F28335 has been presented. The hardware and software implementation of IFOC based IM drive has been described step by step to give the first-time user/student an opportunity to apply this information to specific scenarios rapidly. As presented in this chapter, MATLAB/Simulink environment, has been controled and analyzed the system through a PC on an ezdspF28335 platform. The system has been found to be well suited for learning real-time DSP algorithms for senior level students, graduate students, and practicing engineers.

7.2. Major Contribution of This Book

In this book, two intelligent speed controllers are designed based on sliding-mode and fuzzy logic theory for an indirect field-oriented control (IFOC) of an induction motor (IM) drive. The main contributions of this work are summarized as follows:

- Two intelligent speed controllers based on variable and fixed boundary layer SMC have been developed for an IFOC of IM drive so that the designed intelligent speed controllers have a fast dynamic speed response and they robust to parameter variations and external load disturbances and the chattering is absent in speed and currents under small and large uncertainties.
- The performance of the proposed NFSMC and NBLFC based IM drives have been investigated extensively in simulation at different operating conditions such as load variation and inertia and friction factor variations of the induction motor using some optimized blocks of MATLAB /Simulink.
- The performance of the conventional SMC, FSMC, BLFC, and tuned PI controllers have been also investigated using MATLAB /Simulink. The conventional SMC, FSMC, and BLFC showed superior performance in terms of tracking and transient responses.
- The tuned PI, the conventional BLFC, and the proposed NBLFC, based IFOC of IM drives have been implemented using the Texas Instrument' DSP TMS320F28335 and real-time MATLAB /Simulink. The performance of the proposed NBLFC based IM drive has also been compared with the conventional SMC and BLFC based system.

7.3. Originality of the Work

Author reported some FSMC and BLFC to improve the chattering phenomena and dynamic performance of the IM drive. However, the existing works cannot eliminate the chattering in torque and speed responses of the IM drive completely due to the lack of proper development of those controllers (e.g., linear function inside thin boundary layer of SMC, one input to the fuzzy controller, etc.) [3]. Moreover, in some cases the dynamic performance was not acceptable for high performance drives as the parameters and/or the structure of the controllers were fixed [4].

In the current work first, a novel fuzzy sliding-mode (NFSMC) is developed based on fixed boundary layer SMC, which eliminates the chattering completely from torque and speed responses. In the proposed NFSMC, the fuzzy controller makes a nonlinear function inside the boundary layer to eliminate the steady-state error despite the uncertainties [5, 6]. The applied fuzzy system acts like saturation function technique in thin boundary layer near to sliding surface so that stability of the system is guaranteed. Also, the equivalent control part is estimated to avoid the computational burden by an averaging filter. On the other hand, the averaging filter assists to improve of tracking performance despite the possibility of large uncertainties in the system so that the stability of the system is guaranteed. The main advantages of the proposed free-chattering speed controller are robustness to parameters variation and external load disturbance. Finally a novel boundary layer fuzzy controller (NBLFC) is developed based on variable boundary layer SMC, which improves the steady-state chattering of both torque and speed responses of the IM drive as compared to the existing BLFC [3]. In the proposed NBLFC, the Integral filter assists to eliminate the chattering from torque and speed responses at different operating condition and uncertainties. To ensure the stability of the proposed NBLFC controller based IM drive, the Lyapunov's stability theorem is also applied.

Novel SMC techniques are reported through this work, which have no chattering issue at different uncertainties. Moreover, the proposed NBLFC provides the best dynamic performance as compared to the tuned PI and the existing BLFC controllers. It is verified through simulation and experimental results.

7.4. Future Work

Two types of fuzzy sliding based speed controllers for indirect field-oriented control of induction motor drives are presented in this book. Future works are suggested as follows:

- To reduce the torque ripple, an additional control technique can be developed and incorporated with the proposed technique
- To reduce the cost of implementation, this drive can be implemented without any speed sensor.
- To improve the motor efficiency, a loss-minimization algorithm (LMA) can be cooperated with the proposed drive.
- Since intelligent adaptive nonlinear controllers support a significant amount of all the drives, the proposed intelligent speed controllers can be implemented in other types of drives, but not limited to surface mounted permanent magnet synchronous motors (PMSM), and interior permanent magnet synchronous motors (IPMSM).

REFERENCES

[1] A. Saghafinia, H. W. Ping, and M. N. Uddin, "Fuzzy sliding mode control based on boundary layer theory for chattering-free and robust induction motor drive," *The International Journal of Advanced Manufacturing Technology,* vol. 71, pp. 57-68, 2014.

[2] A. Saghafinia, H. W. Ping, and M. N. Uddin, "Sensored field oriented control of a robust induction motor drive using a novel boundary layer fuzzy controller," *Sensors,* vol. 13, pp. 17025-17056, 2013.

[3] T. Orowska-Kowalska, M. Kaminski, and K. Szabat, "Implementation of a Sliding-Mode Controller With an Integral Function and Fuzzy Gain Value for the Electrical Drive With an Elastic Joint," *Industrial Electronics, IEEE Transactions on,* vol. 57, pp. 1309-1317, 2010.

[4] L. Viet Quoc, C. Han Ho, and J. Jin-Woo, "Fuzzy Sliding Mode Speed Controller for PM Synchronous Motors With a Load Torque Observer," *Power Electronics, IEEE Transactions on,* vol. 27, pp. 1530-1539, 2012.

[5] H. Lee, E. Kim, H. J. Kang, and M. Park, "A new sliding-mode control with fuzzy boundary layer," *Fuzzy Sets and Systems,* vol. 120, pp. 135-143, 2001.

[6] Y. K. Kim and G. J. Jeon, "Error reduction of sliding mode control using sigmoid-type nonlinear interpolation in the boundary layer," *International Journal of Control, and Systems,* vol. 2, pp. 523-529, 2004.

APPENDICES

APPENDIX-A: INDUCTION MOTOR PARAMETERS

Parameters	Value	Parameters	value
Rated power	1000W	Rated torque	3.37N.M
R_s	6Ω	J_r	.0055Kg.m^2
R_r	5.72Ω	B	.001Kg.m^2/s
L_s	428.7e-3H	P	2
L_r	428.7e-3H	Rated speed	2830rpm
L_m	416.6e-3H		

APPENDIX-B: DC GENERATOR PARAMETERS

Parameters	Value
Rated power	1000W
Output Rated Voltage	220V
Output Rated Current	6.4A
Rated speed	2150rpm
Excited Rated Voltage	200V
Excited Rated Voltage	0.2 A

Appendix-C: The Complete Simulink Block Diagram of Closed-Loop Vector Control of IM Drive

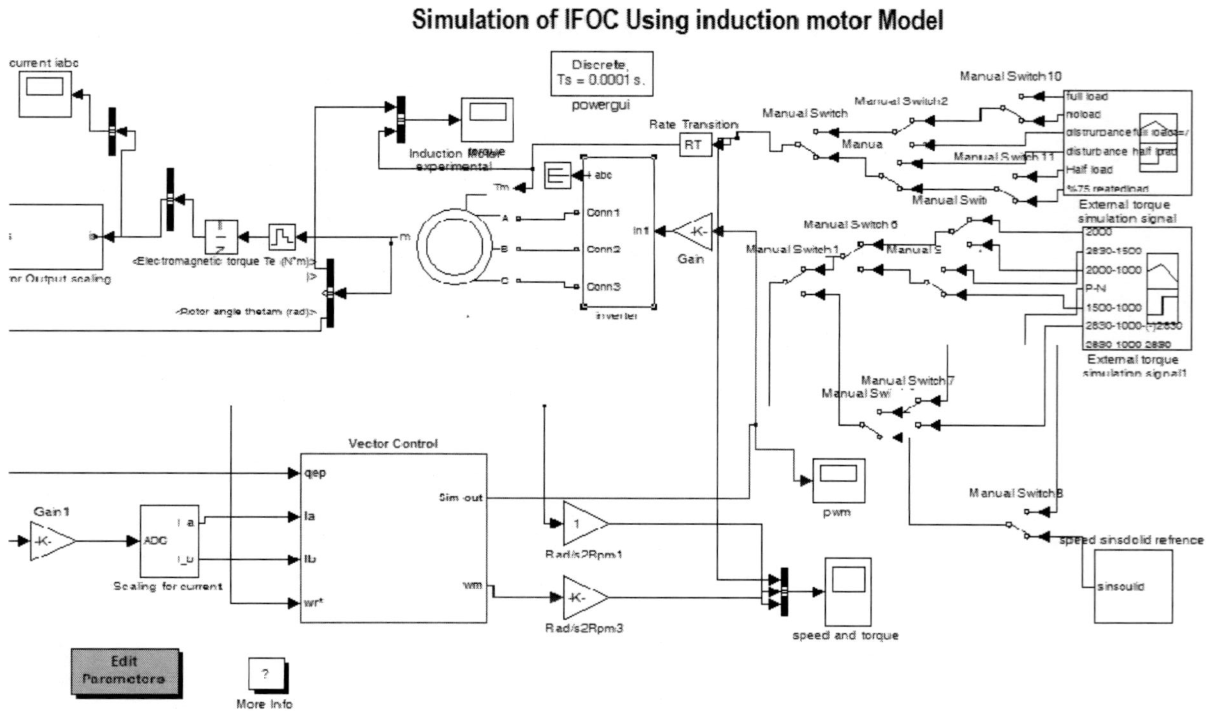

C1. The main Simulink model for the IFOC of IM drive.

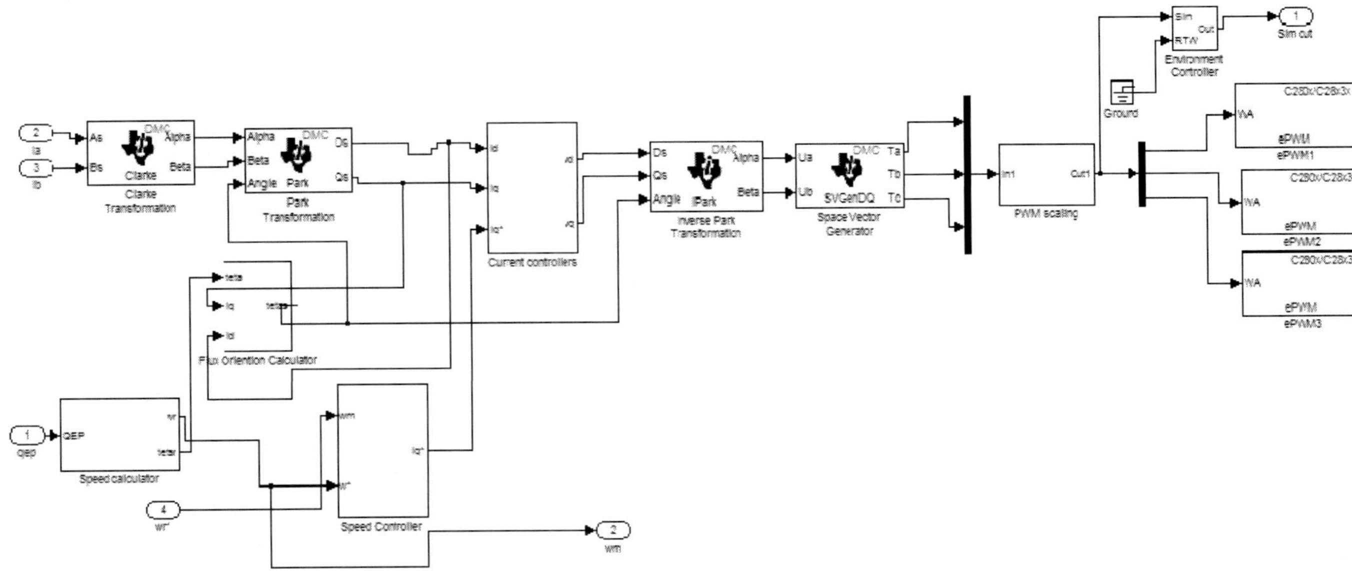

C2. Simulink model of vector control

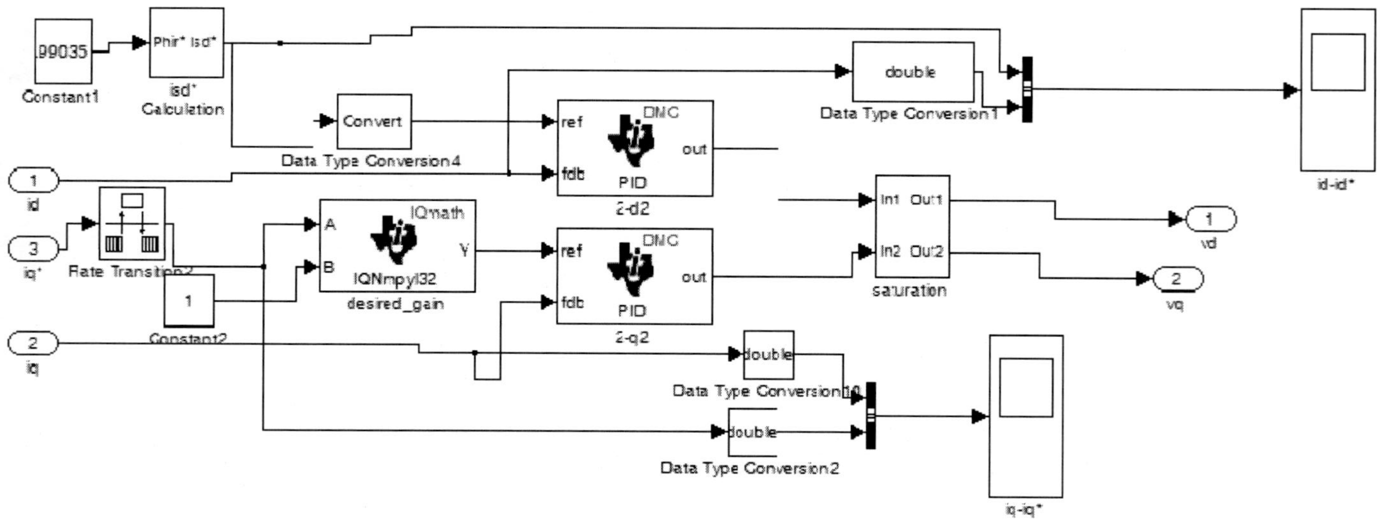

C3. Simulink model of current controllers.

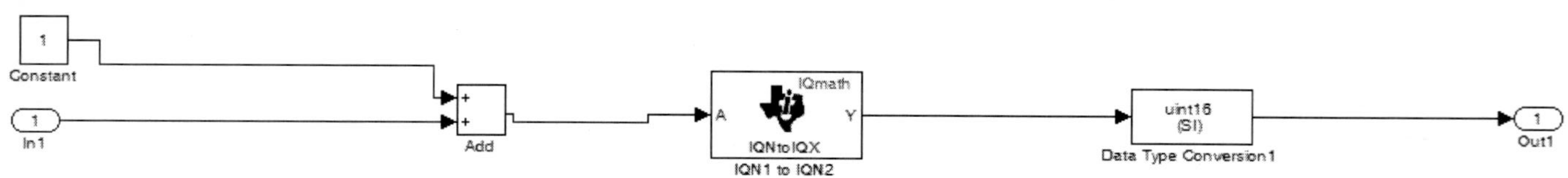

C4. PWM scaling Simulink model.

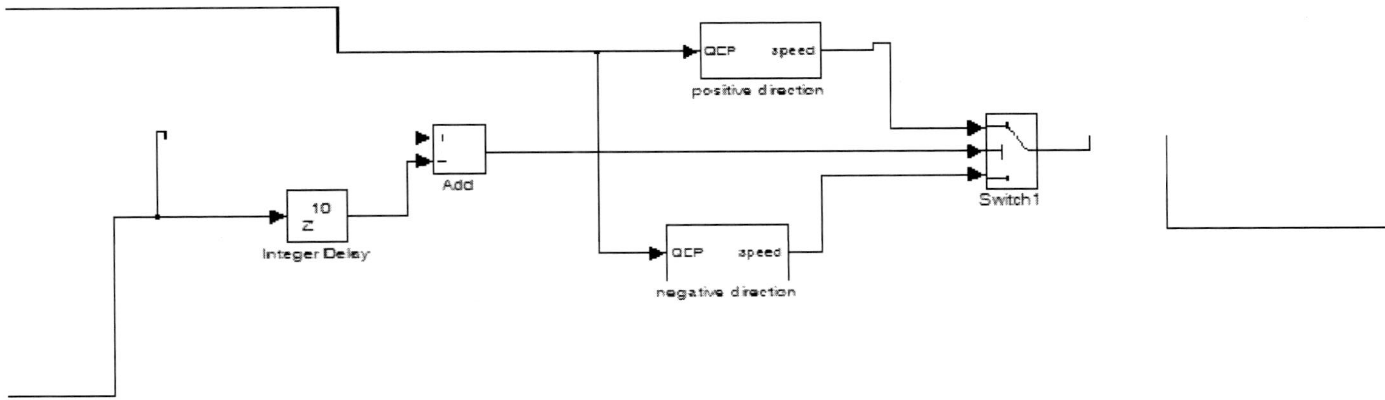

C5. Speed calculator Simulink model.

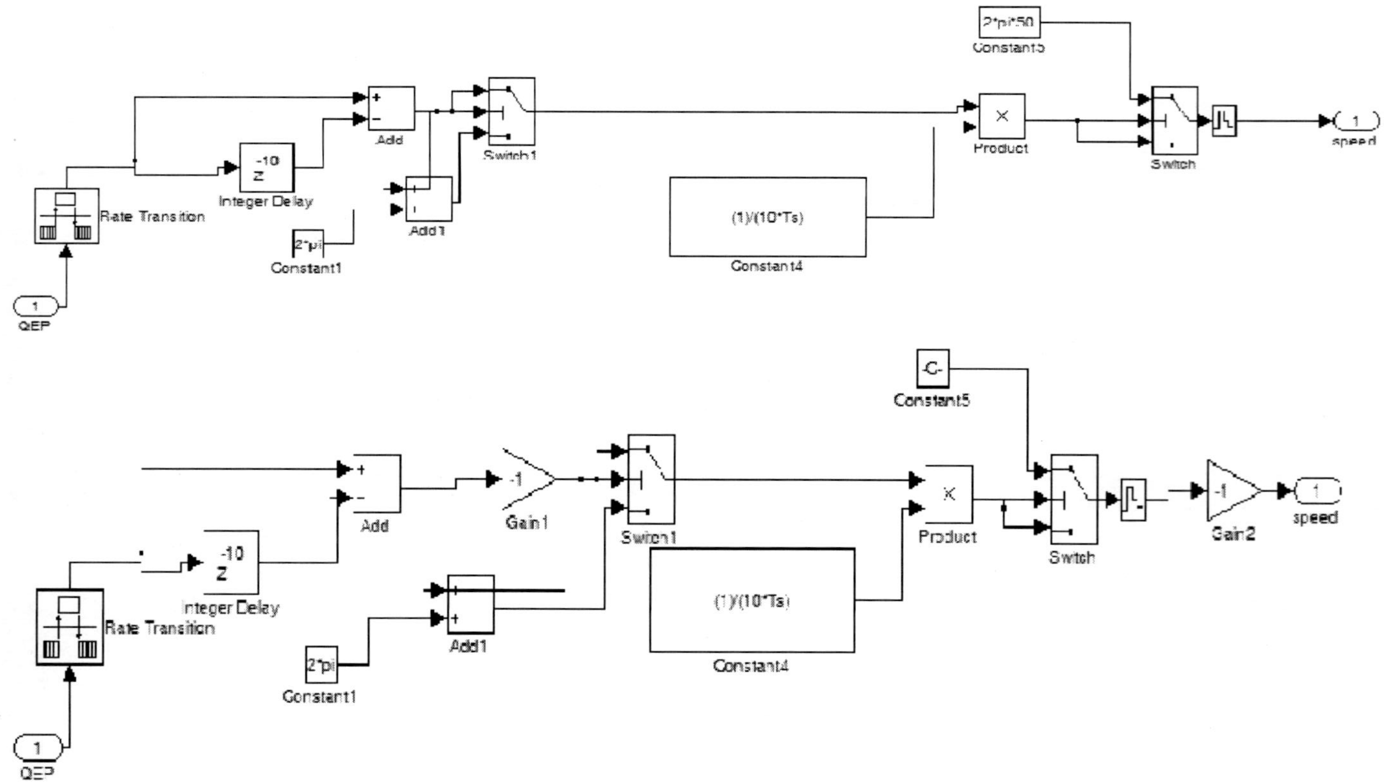

C6. The positive and negative calculation of speed calculator Simulink model.

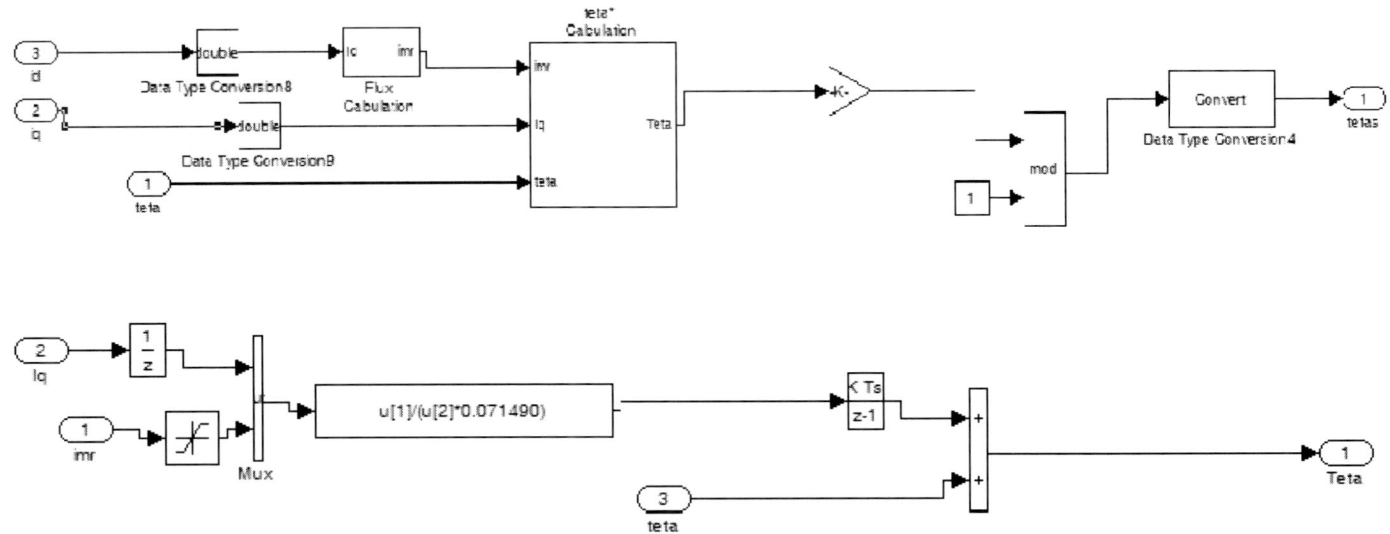

C7. Simulink model of flux orientation calculator.

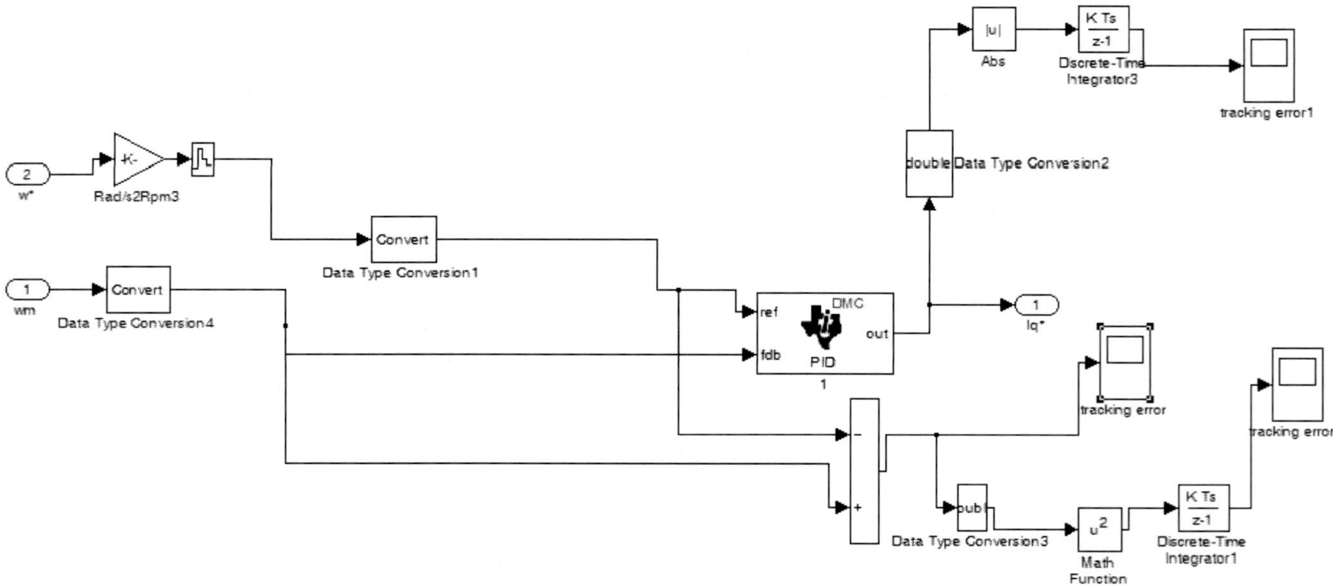

C8. Simulink model of PI speed controller.

Appendix-D: The Simulink Models of the Conventional SMC, and the Conventioal FSMC Speed Controllers

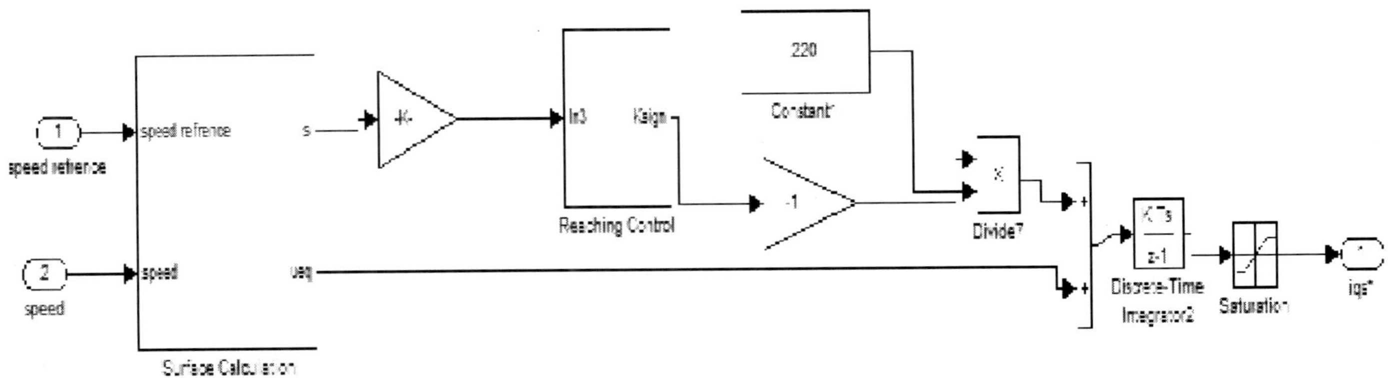

D1. Simulink model of basic block diagram for the SMC controller.

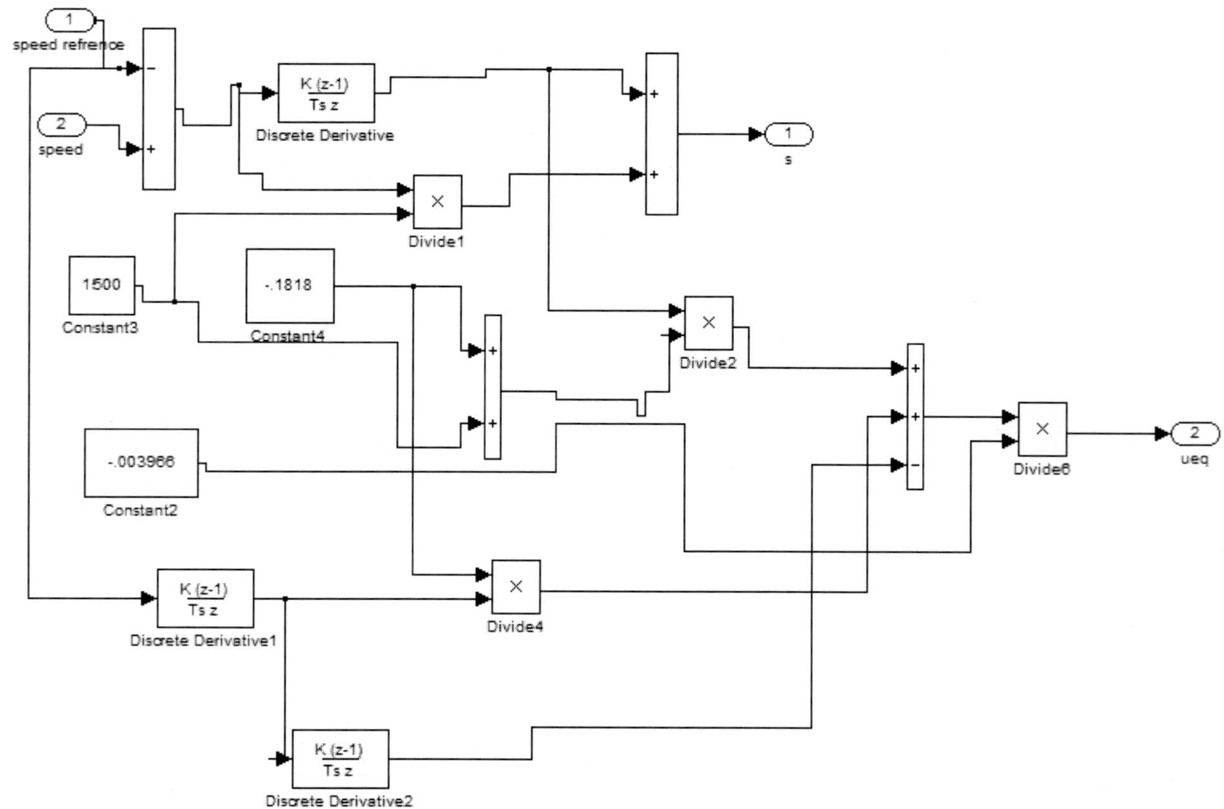

D2. Simulink model of surface and equivalent calculator.

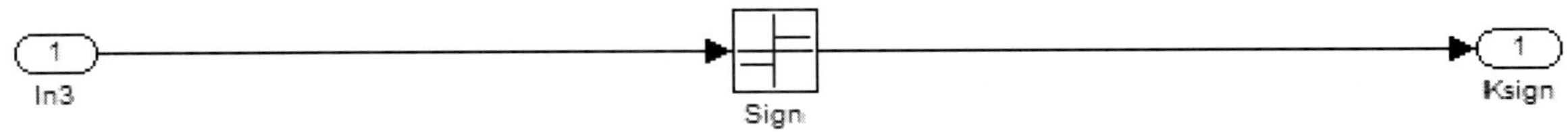

D3. Reaching part Simulink model of the conventional SMC.

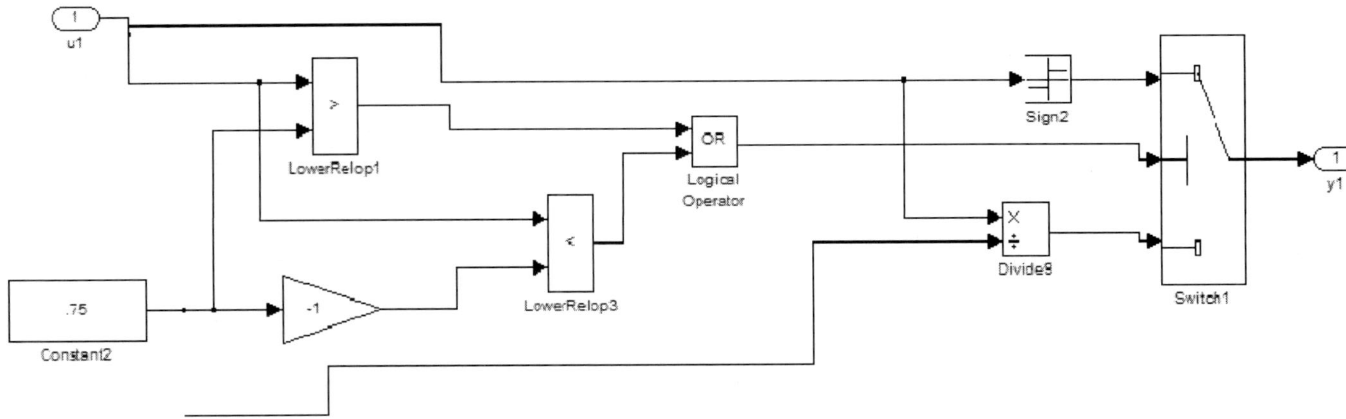

D4. Reaching part Simulink model of the FBLSMC.

Appendix-E: The Simulink Models of the BLFC and the Proposed NBLFC Speed Controllers

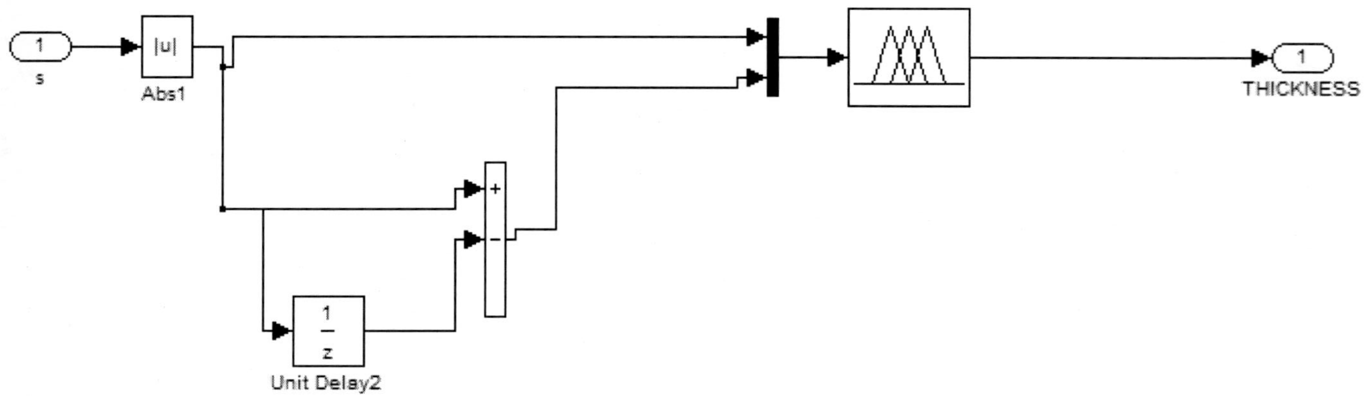

E1. Simulink model for the fuzzy system of BLFC controller.

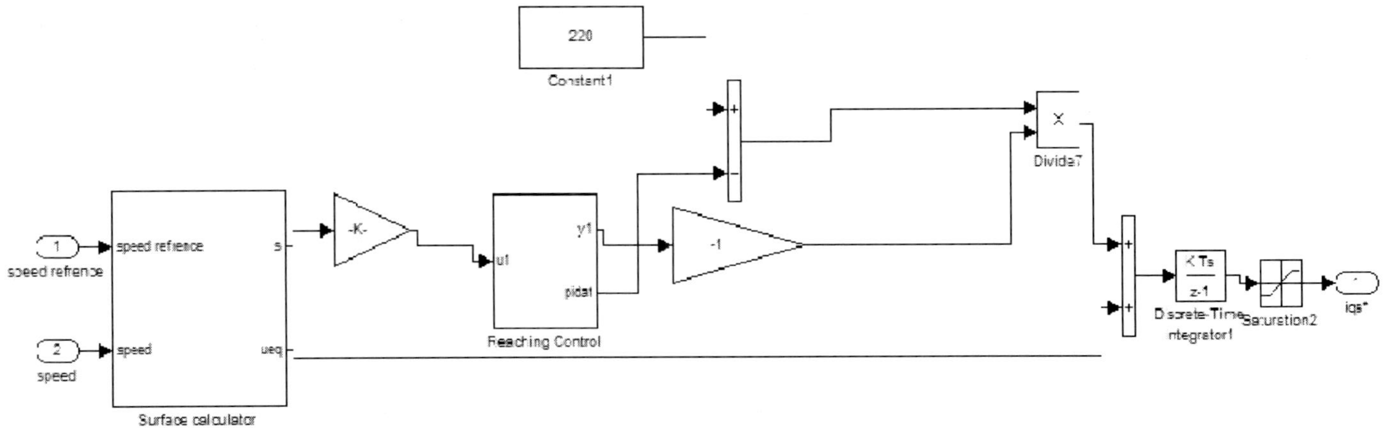

E2. Simulink model of block diagram for the BLFC controller.

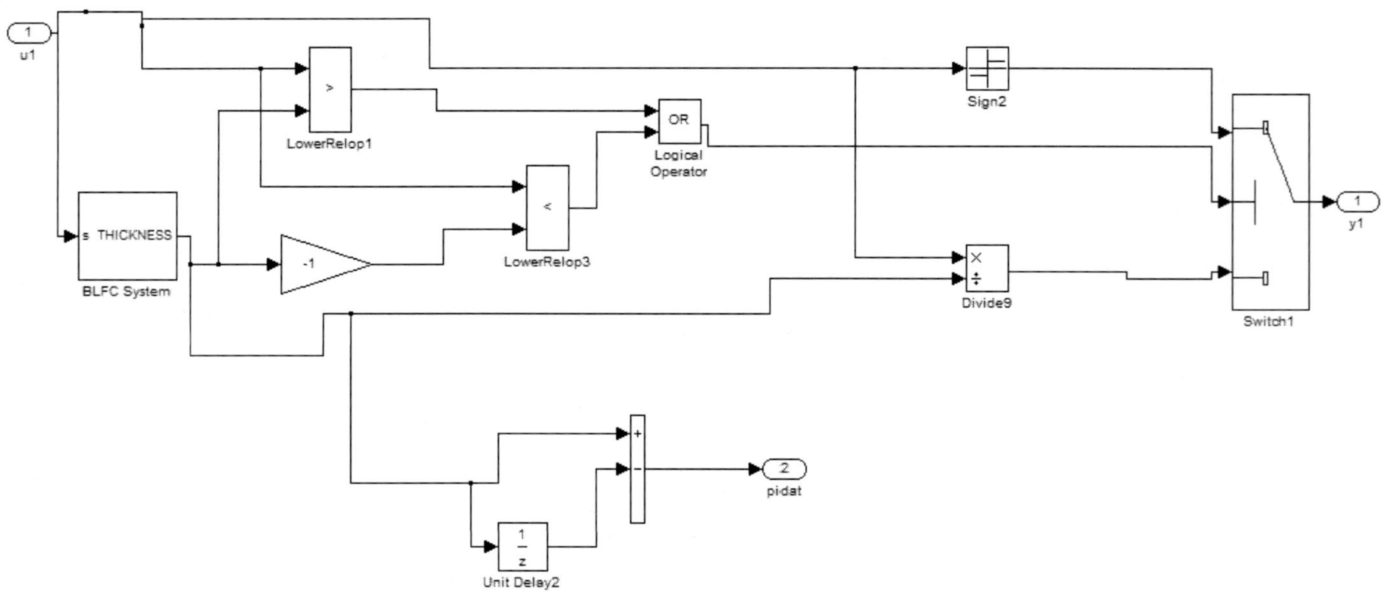

E3. Reaching part Simulink model of the BLFC controller.

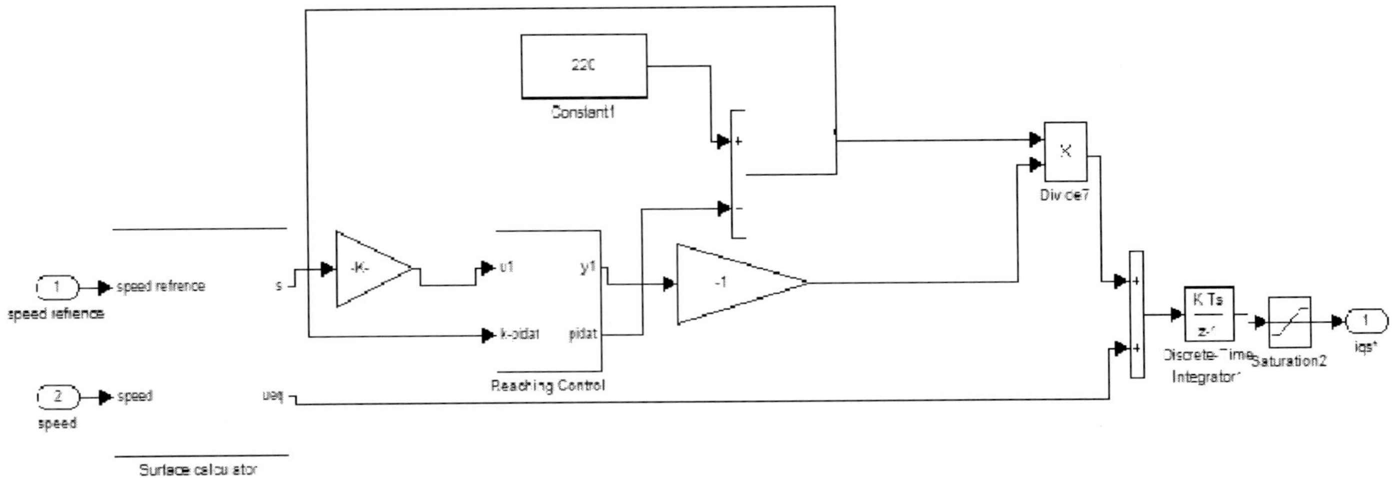

E4. Simulink model of block diagram for the proposed NBLFC controller.

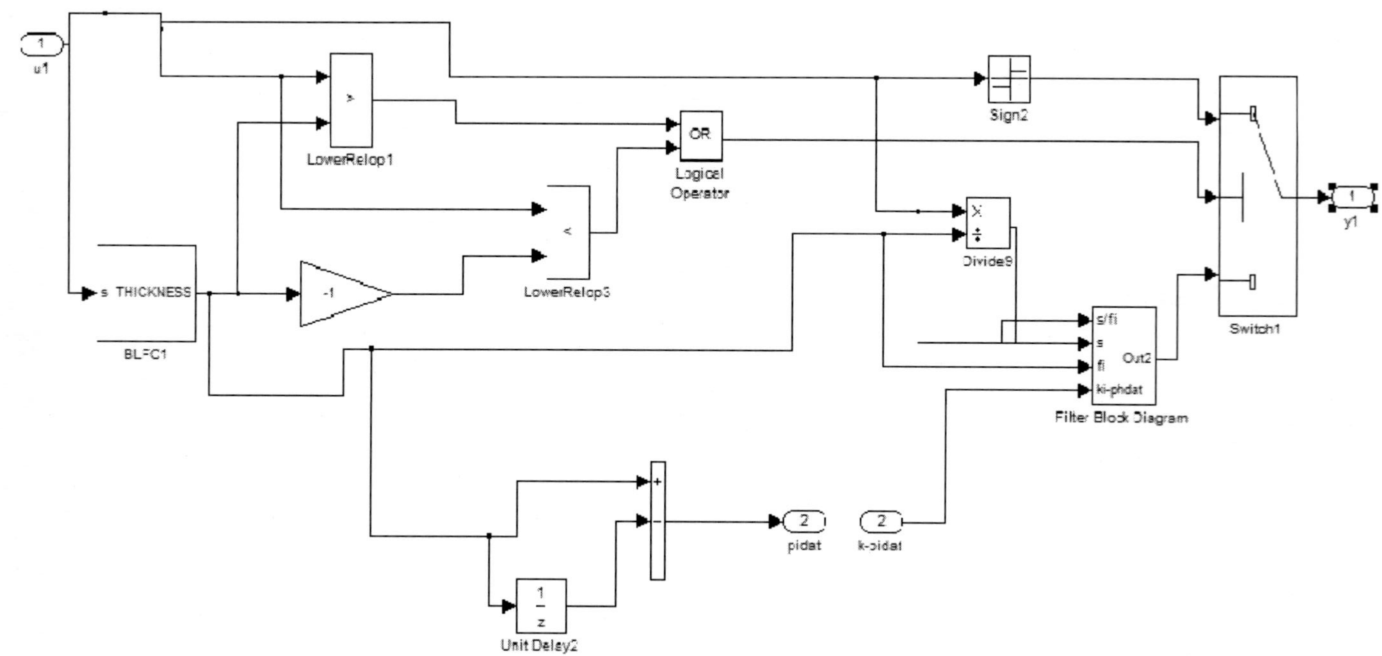

E5. Reaching part Simulink model of the proposed BLFC controller.

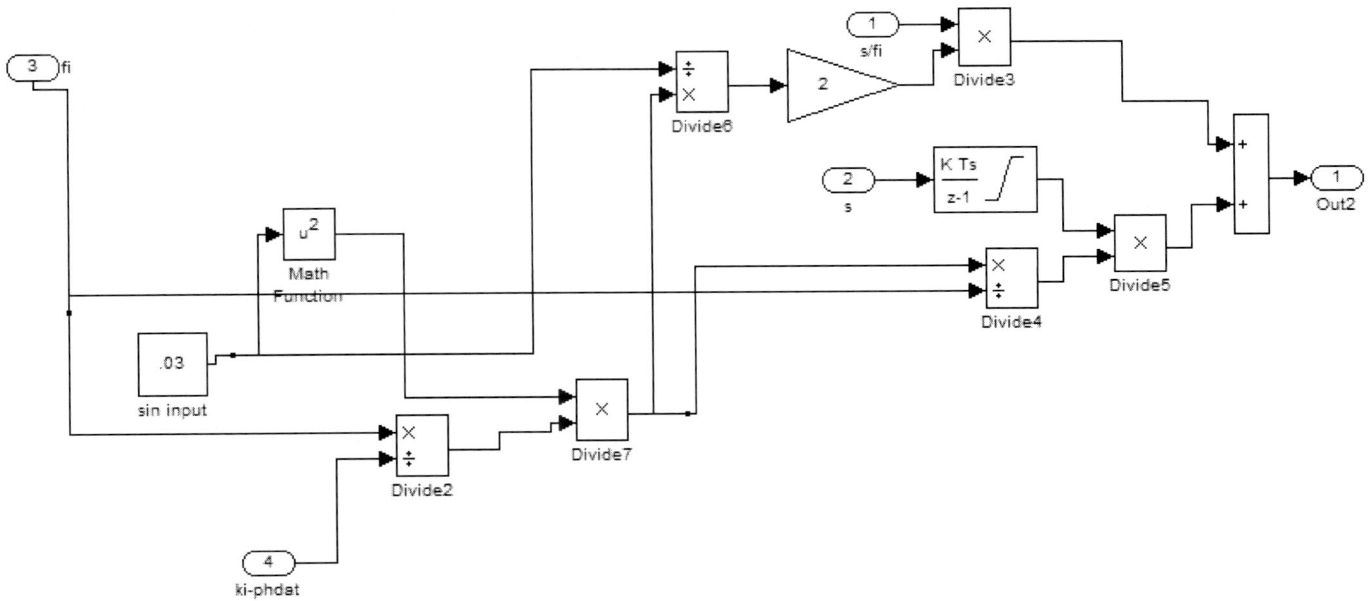

E6. Filter inside boundary Layer for the proposed NBLFC Controller.

APPENDIX-F: THE CHANGED SIMULINK MODELS OF THE CLOSED-LOOP CONTROL MODELS FOR TUNING OF CURRENT CONTROLLER COEFFICIENTS

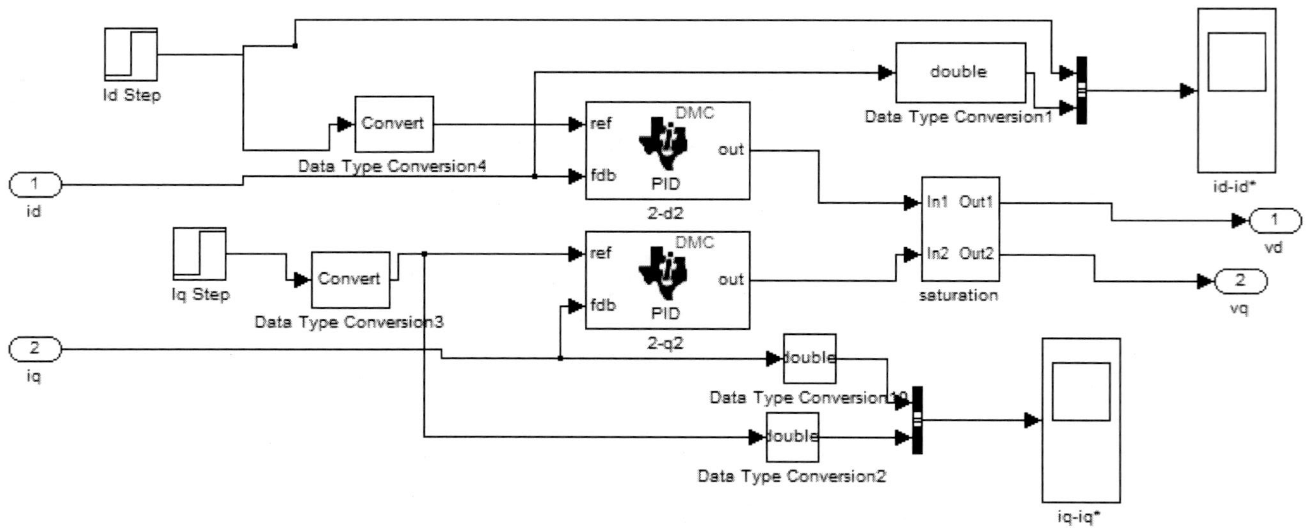

F1. Simulink model of current controllers for the tuning PI controllers.

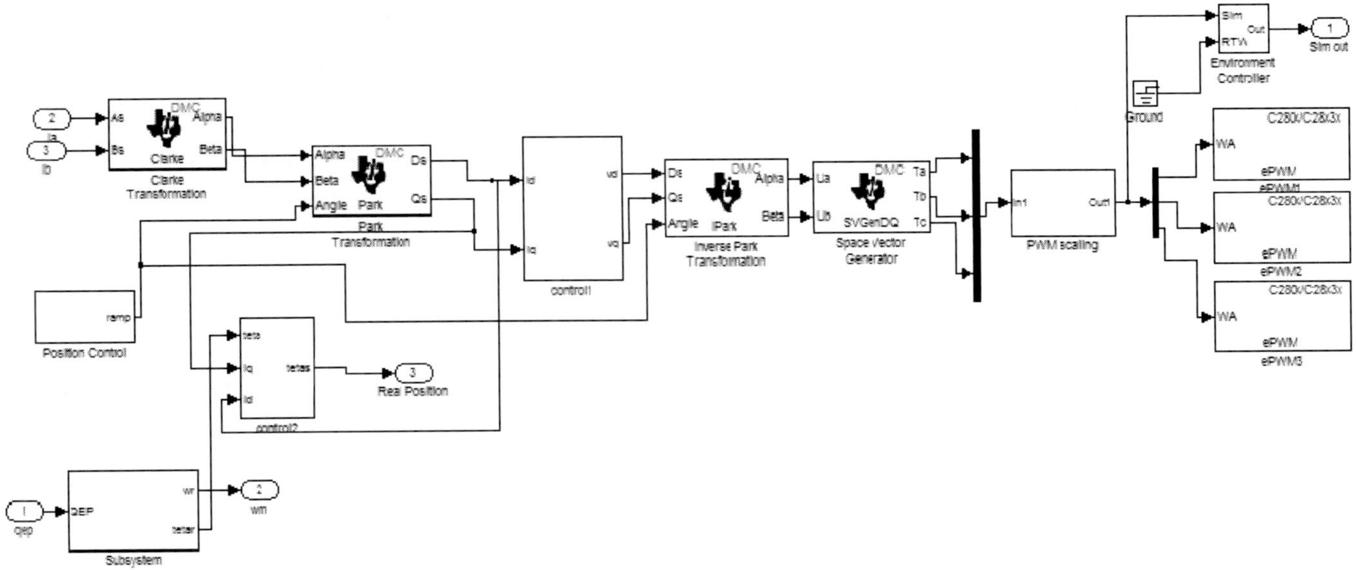

F2. Simulink model of vector control for the tuning PI controllers.

APPENDIX-G: THE DETAIL OF THE DESIGNED INVERTER ALONG WITH SNUBBER CIRCUIT AND THE GATE DRIVE SCHEMATIC

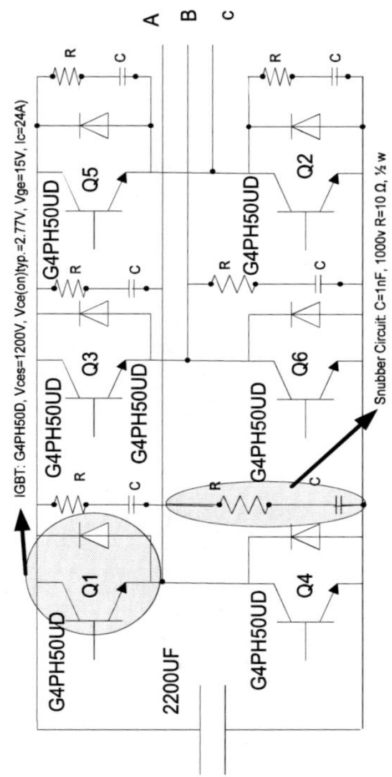

G1. Inverter circuit along with Snubber circuits.

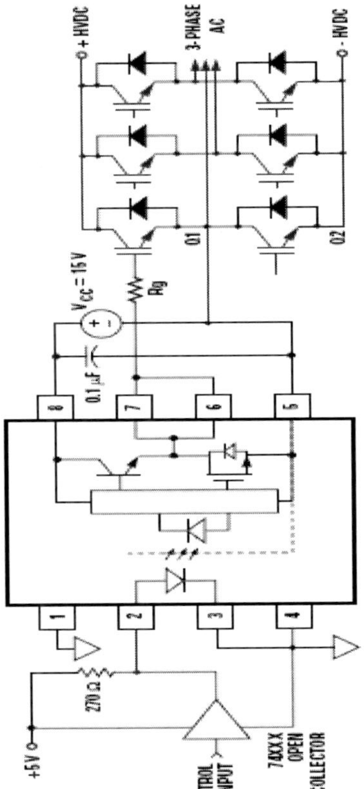

G2. Gate drive schematic for one of IGBT base.

Appendix-H: Real-Time Simulink Model

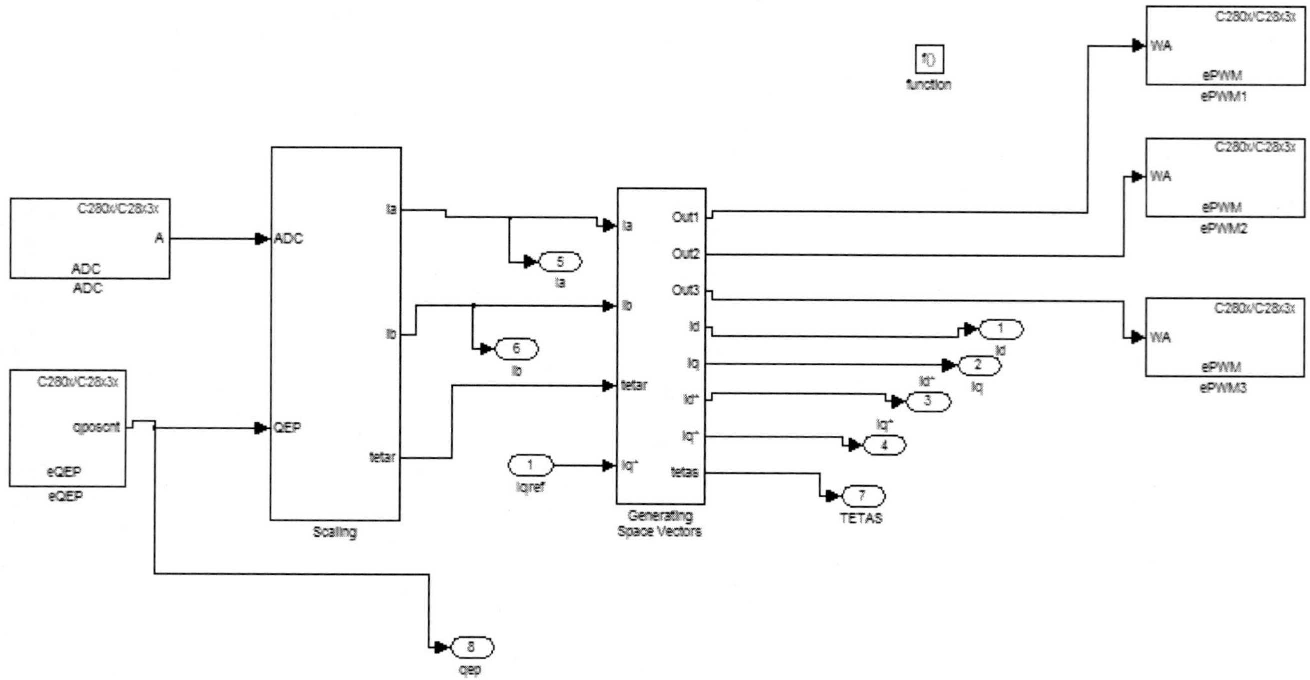

H1. Simulink model of vector control in real-time.

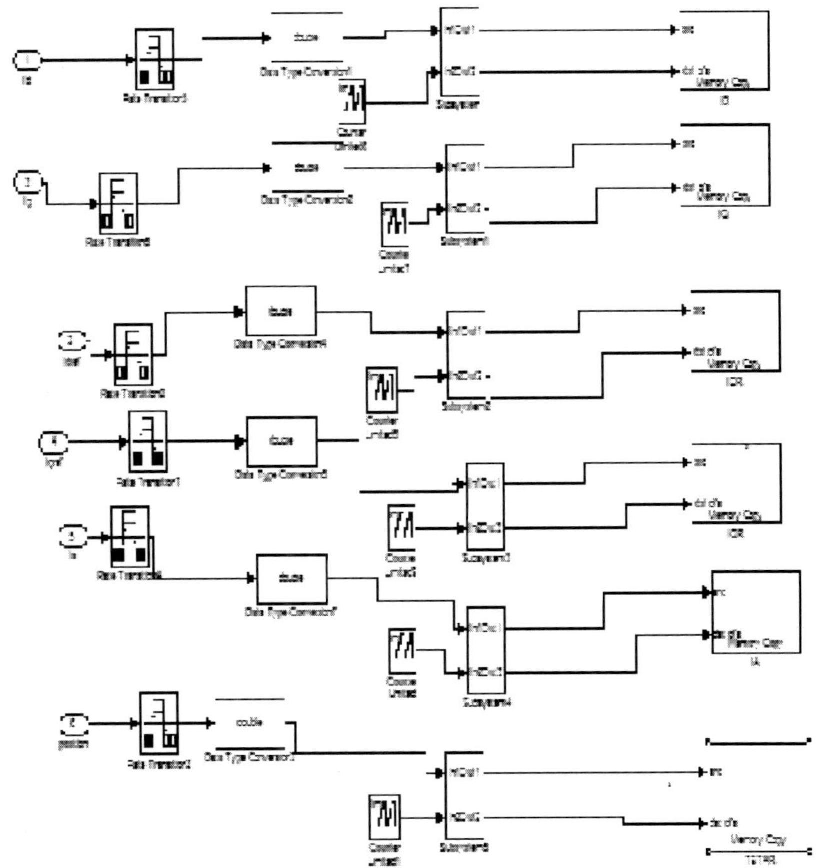
double
Memory Copy
position

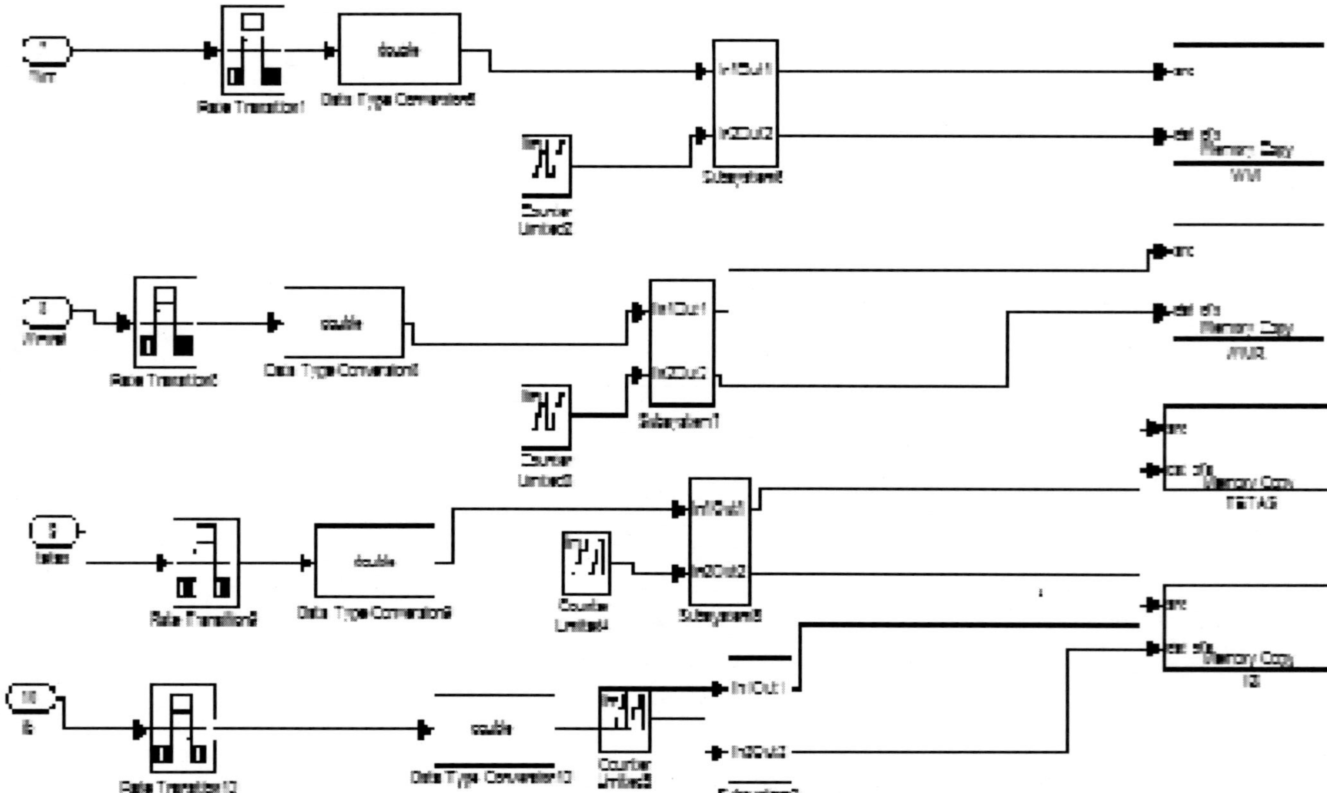

H2. Data saving organization for real-time implementation.

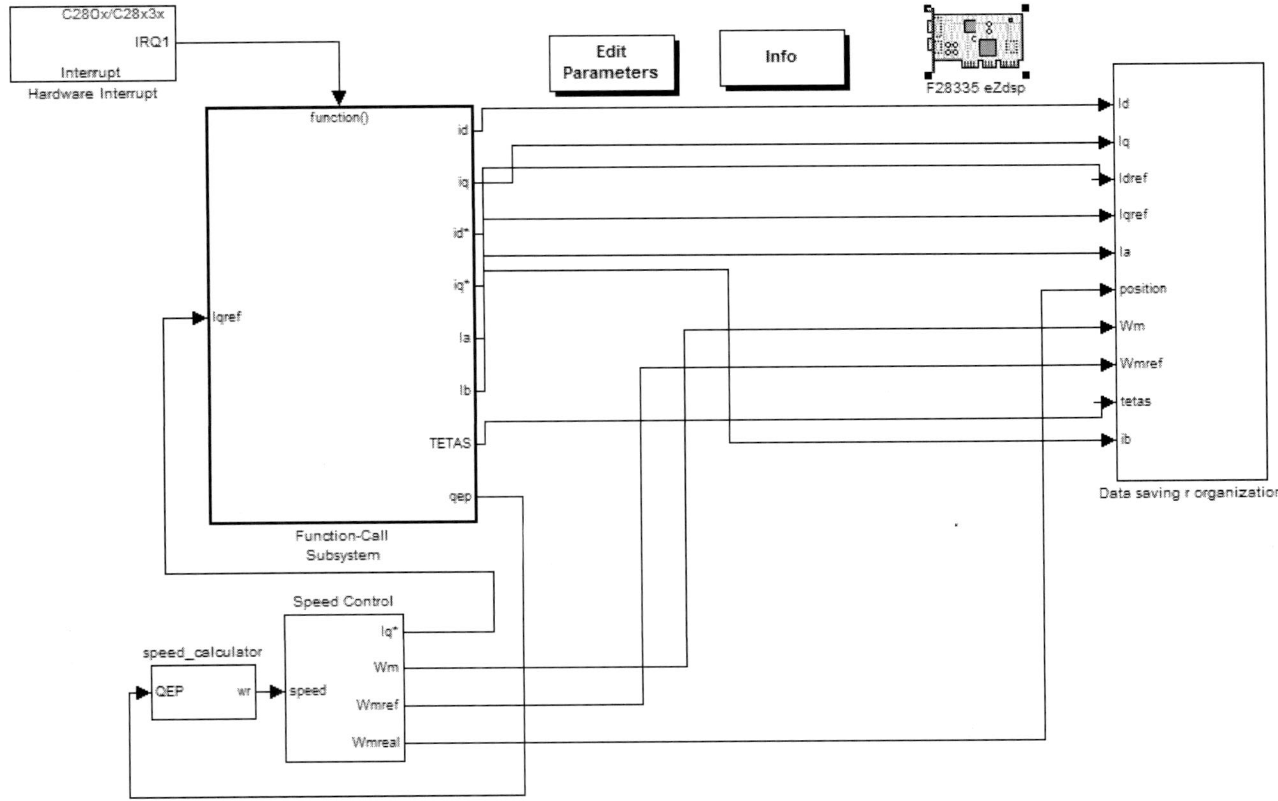

H3. The main Simulink model for the IFOC of IM drive in real- time.

AUTHOR'S CONTACT INFORMATION

Ali Saghafinia, PhD
Assistant Professor
Electrical Engineering Department, Majlesi Branch,
Islamic Azad University,
Esfahan, 8631656451, Iran
Email: Saghafi_ali@yahoo.com

INDEX

L

M

N

O

P

R

S

T

U

V

W